TRENDS IN THE WORLD ALUMINUM INDUSTRY

TRENDS IN THE WORLD ALUMINUM INDUSTRY

BY STERLING BRUBAKER

Published for Resources for the Future, Inc.

By THE JOHNS HOPKINS PRESS, Baltimore

RESOURCES FOR THE FUTURE, INC.

1755 Massachusetts Avenue, N.W., Washington, D.C. 20036

Resources for the Future is a non-profit corporation for research and
education in the development, conservation, and use of natural resources.
It was established in 1952 with the cooperation of the Ford Foundation and
its activities since then have been financed by grants from that Foundation.
Part of the work of Resources for the Future is carried out by its resident
staff, part supported by grants to universities and other non-profit
organizations. Unless otherwise stated, interpretations and conclusions in
RFF publications are those of the authors; the organization takes responsibility
for the selection of significant subjects for study, the competence of the
researchers, and their freedom of inquiry.

This book is one of RFF's studies in energy and mineral resources, which are
directed by Sam H. Schurr. Sterling Brubaker is an RFF research associate.
The manuscript was edited by Elizabeth Reed.

Director of RFF publications, Henry Jarrett; *editor,* Vera W. Dodds;
associate editor, Nora E. Roots.

FOREWORD

In this study of the world aluminum industry, Sterling Brubaker has focussed on the factors that determine the location of facilities for converting raw bauxite into finished metal. The present locational pattern, like that in many other metal industries, finds the mining of the raw ore occurring mainly within the so-called less-developed countries, while the production and processing of metal is located in industrialized countries, where the metal is also mainly marketed.

Within the raw-material-producing, less-developed countries there is a strong desire, fed by both economic and political considerations, to have economic life within their borders reach out beyond the extractive stage of production to encompass as many as possible of the manufacturing operations involved in producing finished metals. In the case of aluminum, unlike other metals, this desire is present not only in the countries producing bauxite (the aluminum ore), but also in those countries in which the potential for low-cost hydroelectric power is thought to exist. A large-scale electricity market such as that provided by an aluminum-reduction works can do much to enhance the economic worth of a multipurpose water-development project.

Certain physical characteristics of the aluminum-production process lend plausibility to the belief that the costs of producing metal and shipping it to distant markets in industrial countries should be favorable at bauxite-producing locations, or at low-cost water-power sites, or at those rare places that combine the two. The production of one ton of aluminum requires four to five tons of bauxite. Obviously, therefore, it would save a great amount of tonnage to transport metal rather than ore. More than is the case in any other widely used metal, the production of aluminum metal

v

is also intensive in its consumption of electricity. Some 17,000 kilowatt hours are required to smelt a ton of metal, so that a difference in the price of electricity of only one mill per kilowatt hour is equivalent to about $17 per ton of aluminum.

Calculations of this sort have aroused the hopes of many in less-developed countries around the world. But although such calculations seem persuasive when looked at in isolation, their significance can be properly appraised only within the complete structure of costs which must be considered when decisions are made concerning location. It is this total picture from mine to metal market for an industry operating on a world scale that Mr. Brubaker fits together in the present study, and from which his conclusions concerning locational prospects are drawn.

Mr. Brubaker has produced an important addition to the existing literature dealing with the economics of the world aluminum industry. The results of his research should prove valuable to those who want an improved understanding of how the far-flung operations of mining, transportation, refining, and marketing of aluminum are tied together on the world stage, as well as to those who are seeking guidance concerning the outlook for future changes in the pattern of location of the industry's physical facilities.

Washington, D.C.   SAM H. SCHURR
October, 1966   Director, Energy and
                                     Mineral Resources Program
                                     Resources for the Future, Inc.

PREFACE

It is difficult, and possibly presumptuous, for one who lacks extensive familiarity with the subject to attempt a study of the world aluminum industry. If one is to have any hope of success in such an endeavor, the cooperation of persons in industry, government, and private research is essential. I have been fortunate in having access to many people whose thoughts have contributed importantly to what follows.

Initial stimulus to the undertaking came from John Krutilla of Resources for the Future, Incorporated. His broad-ranging study and advisory work in the field of water resources made him aware of the frequent tendency to view aluminum plants as a potential market for the power which makes multipurpose river-development projects feasible, and he discerned the need for an over-all appraisal of the industry's prospects if such projects were to be rationally considered.

Within the North American industry, excellent cooperation was received from Tom Covel, George Delany, Odd Kihle, and Fred Lawton, all of Alcan; from Stanley Malcuit, Ken Smith, and Krome George of Alcoa; Irving Lipkowitz and Irving Roberts of Reynolds; John Hendrick and Lafayette Brown of Kaiser; and Milton Herzog of Olin Mathieson. In Europe, Aage Owe of Aardal og Sunndal, Birger Ydstie of Electrokemisk, George Field, R. D. Hamer, and M. B. de Sousa Pernes of Alcan, L. V. Chilton of British Aluminium, Rolf Herklotz-Delaitte of Alusuisse, G. Baudart of Aluminium Français, and Rudolf Escherich of VAW were among the most helpful.

In government and international agencies I sought assistance on matters concerning trade problems and energy prospects. Kenneth Bohr of IBRD, David Feldman of the United Nations, John Styer and Morris Bailkin of the U.S. Department of Commerce, and William Vogely and John Stamper of the Bureau of Mines all provided valuable information, as did Rudolf Regul of ECSC, Georges Brondel of EEC, Christian Laading of OECD, M. de Verteuil of Electricité de France, and Fred Vogt and G. Roald of the Norwegian Watercourse and Electricity Board.

In addition I have profited from exchanges on this subject with Sam Moment, an American consultant, Professor Hans Bachmann of St. Gallen, and W. F. Wildschutz and L. Ernst of Aluminium Zentrale. A special mention must be made of the generosity of Metallgesellschaft and of Mr. Wiessler of that organization who provided basic statistics on production, consumption, and prices.

Within RFF the interest and comments of Sam Schurr have been of great value, and excellent critical review was received from O. C. Herfindahl. Finally, Sally Nishiyama was an energetic and skillful research assistant who contributed in a variety of ways to the task.

Because of the great divergency of viewpoint and interest of persons consulted, it would be impossible to satisfy all of them either from the standpoint of approach or conclusions. Thus, while thanking all who aided me, I must claim responsibility for the final product.

October 3, 1966   STERLING BRUBAKER

CONTENTS

LIST OF TABLES

xiii

TRENDS IN THE WORLD ALUMINUM INDUSTRY

CHAPTER ONE

INTRODUCTION

When primitive man stirred the ashes of his cooking fire, he found there iron, copper, and zinc from which he fashioned articles of utility and war. He found no aluminum, although the metal trickled through his fingers in concealed form in almost any handful of earth he might pick up. A modern metal, aluminum was not to be recovered by accident but only by the methodical application of then-undiscovered principles of science.

Intrigued by early bits of the metal he had come across, Napoleon III had them made into his most exclusive table service and dreamed of the day when the metal might be available for military applications. That day came by the end of the nineteenth century and his dream was fully realized with the advent of the aerospace age; from the beginning aluminum has carried the stamp of a strategic military material. The metal's use for exclusive table service, however, has passed long since. Today it is so commonplace that man may replace his cooking fire with an aluminum solar cooker on which he is very likely to heat his cheapest pots—of aluminum.

Nonetheless, within the lifetime of many people now living, aluminum has gradually penetrated the consciousness of industrial societies. If steel was the workhorse of the industrial revolution, the light metal has been the queen of a newer technology. Aluminum has bridged the gap from railroads to rocket ships. Although it is tempting to recount the elements of excitement and glamour in the story of aluminum, it is unwise to lay too much stress on the more exotic aspects of an industry which, by force of habit, we still refer to as young. In fact, the aluminum industry is approaching middle age and there is need, therefore, to look more realistically at the question of how it will settle into the future world economy.

The association of aluminum with modern technology and industrial and military strength has commended it to the attention of planners in many ambitious new countries, especially those which may have important bauxite or hydroelectric power resources. In combination with a somewhat oversimplified view of the prerequisites for producing and marketing aluminum competitively, this has given birth to aspirations that may not be entirely realistic.

Since heavy investment outlays are required to support this capital-intensive industry, it is all the more imperative that less-developed countries avoid mistakes in the timing and location of investment in aluminum facilities. The objective of this study will be to examine some of the factors at work that will affect the future size and location of the world's aluminum industry. This will require attention to the industry's structure and to the importance of the trade and investment climate quite as much as to technology and costs. There is no attempt to make specific location studies of a sort which firms would require in siting a plant or which a government agency would need to attract investors.

Prior to 1940, virtually all of the world's aluminum was produced in the industrial countries of Europe and North America and in Japan. Today these areas still predominate, but the pattern is less monolithic. One may ask how the earlier pattern was established and what changes have occurred to bring new locations into production. No single answer will suffice, but the main ingredients can be discerned.

The locational rationale up to World War II can be characterized briefly. As the product of an advanced technology long protected by patents and other devices, aluminum was destined to be tightly held by the handful of firms who had access to this technology. These firms were often national monopolies and were not under competitive pressure to seek lowest-cost production sites on an international basis. They preferred to operate in the familiar home institutional environment near their markets, protected by tariffs and benefiting from an industrial environment. A production venture abroad might mean moving into a nearby advanced country such as Canada or Norway. Moreover, their cost structure generally did not compel them to look abroad, since, at the then smaller scale of the industry, bauxite resources in Europe and the United States were able to meet a major part of local needs and

4

inexpensive hydroelectricity could be had by locating in the more remote areas of industrial countries. Many of today's underdeveloped countries were that day's colonies and were unable to pursue independent policies of industrialization that might have stimulated relocation of the industry. Just as today, the absence of infrastructure and industrial skills limited their attractiveness to the industry. Finally, the ever-present concern with aluminum as a defense material led to overt state support of the industry in Japan and Germany, and tariffs or commonwealth preferences fostered French, United States, and Canadian production.

A number of factors—some of continuing effect—have altered this pattern and other changes are apparent that will operate in the years ahead. The industry today has a quite different shape from pre-World War II days and it cannot be complacently assumed that the existing pattern will provide the mold for the industry of the future. Moreover, the prospect that the industry will double or triple in size by 1980 offers wide scope for the play of new locational influences.

What are some of the factors which create the possibility of major shifts in the location of production? An obvious one is the change in sources of bauxite supply. Although France remains an important producer of bauxite, all of the other major aluminum producers now are dependent on outside supplies. There is no near-term problem of the global adequacy of bauxite supplies, but known reserves tend to be located in less-industrialized countries. This adds a new element to the location equation to the extent that such countries press for more local processing prior to export of their raw materials; in addition, the shift in bauxite source alters the transport equation for the industry.

Technological changes in the industry also must be considered. Improved technology affects the industry's power requirements and its inputs of capital, labor, and raw materials, as well as its logistics. Such developments may have conflicting implications for location.

The changing size and location of the market for aluminum (in combination with other influences) are major influences on location. Growth facilitates locational shifts, for it means that they can be accomplished without abandoning existing plants. Further, as more national markets attain the size that would support an aluminum smelter, the question of whether to produce their metal

5

domestically becomes real for the first time. Growing population, income, and wider industrial application of aluminum offer this possibility in an increasing number of countries. Meanwhile the growth of markets within industrial countries may outstrip the previous resource base (especially the low-cost power resources) of their domestic industries.

At the same time, the structure of the international industry changes. National monopolies have tended to break down and a growing cross-penetration of markets has occurred. The North American firms emerged from World War II increased in number and in size. For a time their energies were occupied by the problems of postwar adjustment and by expansion during the Korean War, but thereafter they turned their attention and their enormous resources to the outside world. In a climate of improved international liquidity and diminished barriers to trade they have penetrated European and other markets. Major European firms have recognized that the changed environment requires greater flexibility on their part and they too have left their national redoubts. The resulting greater awareness of and sensitivity to possibilities outside their original home countries represent a significant change in the climate of the industry. It is less and less appropriate to identify the major firms by a national label.

Perhaps what has occasioned most speculation about the future of the world aluminum industry is the concern that there will be insufficient low-cost electric energy available in developed countries in contrast with the possibilities in underdeveloped countries. Power cost is a major locational factor in the aluminum industry. Growing industrial and residential demand for power in developed countries has brought most of the inexpensive hydroelectric sites into production. Although in peripheral areas such as Canada and Norway this process has not been completed, even there long-distance transmission offers the possibility of an alternative market for power which otherwise might find the aluminum industry its best outlet. Meanwhile, the cost of conventional thermal energy has remained discouragingly high in Europe and Japan. With aluminum consumption still growing rapidly and with power expensive in many industrial countries, attention shifts to those underdeveloped parts of the world where power could be produced cheaply and where no competing demands would bid up its price.

Such a glance toward less-developed countries encounters ready interest there. Newly independent countries, anxious to industrialize, view heavy industries such as steel and aluminum as symbols of progress. Such industries also may hold a very practical attraction as potential foreign exchange earners of a sort different from the all-too-vulnerable traditional exports of such countries. Moreover, it sometimes is hoped that aluminum, as a great consumer of power, will provide a market and cash return to multiple-purpose river development projects—a market on a scale that no other industry could offer and which may be essential to the fulfillment of development plans of wider scope. Thus the aspirations of less-developed countries appear to dovetail nicely with the needs of industrial countries in a situation where other forces are acting to break the previous pattern of location.

Having set the industry adrift in this manner, however, it is necessary to examine these elements more closely to see whether their potential import for relocation of the industry is likely to be realized in fact. If this is done, it seems more probable that, while some movement will occur in response to these considerations, there are strong forces at work that are likely to preserve the position of the traditional producing centers.

Any such judgment must be qualified and confined in time. The major basis for it and the chief qualifications will be previewed in the following paragraphs. The horizon chosen for this study is the period out to 1980, since it should be possible to bring to fruition by that date the major hydroelectric projects that may be initiated within the next few years. Beyond that time, the future becomes too hazy to affect current decisions and new studies will in any case precede more remote decisions. Fortunately for the industry, its continuing growth provides shelter from some of the consequences of investment mistakes that otherwise might be serious, and many interim adjustments to the situation are possible as it evolves over the period of our concern.

Let us look, then, at the likely effects of some of the changes occurring in the industry. The growing reliance on tropical sources of bauxite might be thought to give a locational advantage to bauxite-supplying countries. However, few bauxite-supplying countries are exceptionally endowed with the electrical-energy potential essential for smelting. Moreover, though weight loss from ore to metal is on the order of 4 or 5 to 1, the transportation advantage does not lie with smelting near the ore. It lies rather with smelting near to metal markets following the intermediate production of alumina near the ore source. This transportation equation is grounded in the real economies to be had in moving bulk commodities and is not likely to be upset by changes in the competitive situation in shipping. Finally, the abundance of the ore, its distribution in a number of countries, the discovery of major supplies in Australia, and the technical feasibility of using somewhat inferior domestic substitutes all act to limit the amount of leverage that bauxite suppliers in less-developed countries have on the industry.

Technical changes occurring in the industry will permit some marginal expansion at existing sites at low cost. Beyond this, other possible changes may reduce capital costs and foster smaller-scale operations by omitting the intermediate alumina stage. Such a development would act in several ways to favor locations outside the existing centers of the industry—provided the host countries could reach agreements with owners of the new processes. So far, however, the adoption of such new technology remains only a subject for speculation. Even if adopted, the new processes would not alone be sufficient to induce a major locational shift of the industry.

Industrial countries will continue to provide the major markets for aluminum by 1980, although significant markets will be found in India and Latin America as well. Markets influence production locations, both through the favor commonly extended by governments to domestic producers and because of transport considerations. The newer markets may aspire to meet some or all of their needs through local production, especially if they are prepared to welcome outside investment and to offer protection to metal from domestic plants. Whether established industrial-country markets meet their needs for additional metal by domestic production will depend upon costs and conditions of trade, but under foreseeable circumstances local production is likely to remain competitive or to become even more so.

The changed structure of the aluminum industry opens up the location question because of the decline of national monopolies which thereby gives an opportunity to less-developed countries to become exporters. However, less-developed countries face major obstacles if they seek to become independent exporters. Because vertical integration is the practice in the industry and is likely to become still more prevalent at the fabricating stage, would-be exporters to industrial countries are likely to need an affiliation with a major international firm. Control of the industry will continue to reside with these organizations, particularly in international trade. Such firms, released from the narrow perspectives of their national origins, are prepared to produce and sell wherever it proves most advantageous for them, but at the same time they are profit-making institutions alert to the constraints and hazards involved in establishing plants beyond the confines of the legal and institutional framework of Western societies. Therefore, if they are to consider smelter locations in less-developed countries aimed at industrial country markets, they will expect to find sufficient cost advantages to offset the tariffs they will encounter and the added risk as reflected in a higher expected rate of return.

The presumed cost advantage of underdeveloped countries depends mostly on the possibilities of low-cost hydroelectricity to be had there in contrast with the exhaustion of such possibilities in the industrial countries that increasingly must rely upon higher-cost thermal power. However, continuing economies in the use of electric power diminish its importance as an element of cost. Meanwhile conventional thermal power can be produced in the United States at costs that are at no critical disadvantage to the best hydro costs in less-developed countries. Also, nuclear power now can be made available to any industrial country at a cost near to that for the cheapest conventional thermal power in the United States. While less-developed countries may continue to have some advantage in power cost over industrial countries, it is not likely to be decisive for location because the ceiling imposed by nuclear power allows only a narrow margin of advantage that is insufficient to overcome other disadvantages in trading barriers and investment climate.

At the level of supplying their domestic markets, developing countries may aspire to build an aluminum industry. This can be done once their market attains a minimum size. The cost of producing locally will depend upon scale and resources but, if favored by government protection, the industry can survive and ordinarily it will allow some savings in foreign exchange. If the country aspires to be a major exporter, however, it then faces obstacles of the sort indicated earlier.

9

Since there will remain a real (though narrow) advantage on power cost during the period of concern to location in less-developed countries, their export prospects would be much enhanced by any developments acting to reduce trade barriers or to dissolve the investment disadvantages from which they presently suffer. One can only speculate concerning these matters.

Several tendencies affecting trade could benefit the less-developed countries. A general decline in tariffs via the Kennedy Round is still a possibility. The formation of regional trading groups among less-developed countries, of which there is some possibility, would permit more economical provision of their own needs. Affiliation with the trading blocs of developed countries will be possible in some cases. Positive discrimination favoring the industrial products of less-developed countries in developed-country markets has been sought by some.

As far as investment in less-developed countries is concerned, some industrial countries are prepared to offer incentives in the form of tax breaks or investment guarantees. The International Bank for Reconstruction and Development (IBRD) recently has proposed offering loans to such countries at lower rates than in industrial nations. There is increasing discussion of multilateral assistance programs which would make investment funds available to less-developed countries. As another example, a United Nations program of several years' standing in the Mekong Valley could acquire new vitality in response to the political needs of major powers involved there.

Thus, in some circumstances, the obstacles now retarding development of the aluminum industry in less-developed countries might be overcome. It is pointless to speculate on these possibilities. Our conclusion must be that with the existing and foreseeable distribution of markets, with the present or most likely climate for trade and investment, and with prospective relative cost positions, the primary aluminum industry is most apt to remain heavily concentrated in major industrial countries.

Our focus in what follows is on the question of prospective location of the primary aluminum industry. Although this would appear to be a legitimate question under any circumstances, it is especially appropriate now because of changing influences at work on the industry, because of the great absolute growth which may be anticipated over the next 15 years, and because of the aspirations of less-developed countries to become producers and exporters.

Since our view is a prospective one it is insufficient to look merely at present cost and market relationships. Discussion rather must be permeated by concern with how they are likely to alter over the period under review. Inevitably this forces us to depart from the comfort of neat and comprehensive figures (such as might be arrayed for a current location decision) in the direction of a somewhat more qualitative approach. However, in our judgment it is possible on this basis to discern the most likely general direction that the industry will take. Furthermore, in view of the forces apparently determining that direction, a more detailed exposition of future location seems superfluous.

A location decision, to the extent that market forces control, is determined by the costs of assembling materials, processing, and transporting to market. The preferred location is that which permits delivery in the market at lowest cost. In the real world such decisions are complicated by barriers to trade and may also be affected by considerations of corporate strategy and by national interest as interpreted by governments.

Our first problem is to locate the markets to be served and to make some estimate of their future size. This is done in Chapter 3 after a review of the historical consumption and production trends in Chapter 2. There is no attempt at a definitive forecast of future demand. The purpose is merely to arrive at plausible and approximate magnitudes which will give some perspective to future discussion of the location problem. The conclusion that consumption of aluminum will continue to grow rapidly and will be concentrated in industrial countries seems unassailable—regardless of how one feels about marginal shifts among these countries or between them and less-developed countries.

The cost of producing aluminum depends upon the quantities of inputs required and their prices. To understand these possibilities requires an examination of the technological characteristics of the industry which are reviewed in Chapter 4. As with nearly all industrial processes, there is some range of possible variation in physical inputs, with the choice between them depending on their prices, but the process inherently is capital intensive, requires large amounts of electric power, and must handle a great bulk of materials along the way. Thus, investment, transport, and power are key inputs whose provision at low cost becomes important to the quest for lowest-cost production sites. Since exceptionally favorable situations with respect to one or more of the key inputs may permit higher costs on others, it often is misleading to speak of a typical cost structure for the industry. Discussion of typical ranges of input costs is found in Chapters 7 and 9. Because major technological changes in the industry may occur and thereby sharply change the pattern of inputs, some discussion of this possibility has been included in Chapter 8.

Location is not determined merely by these real economic considerations. The enterprises operating in the industry take a very long view of their interests. Where a production location is a means of securing a domestic market, they frequently are willing to locate in areas which, under conditions of free trade, would be more cheaply served from elsewhere. If tariffs exist, they are a part of the cost structure to the outside supplier and he must consider whether to pay them or to locate production inside the market he would serve. At the same time, major international firms are very anxious to maintain their positions despite their widely varied resource, market and capital positions, and their national connections. In addition, very often national governments are prepared to offer incentives or protection to local production. The pattern of trade which has evolved under these influences and the effect of company and national objectives on location are discussed in Chapters 5 and 6.

The final chapter (10) is an effort to bring these economic and quasi-economic elements together. Here, as throughout, the interest centers on the prospects for developed as against less-developed countries as sites for smelters aimed at developed-country markets. A decade or more ago industrial countries became conscious of their lack of low-cost power to meet the needs of an expanding industry and there was common belief that the industry would turn in the direction of less-developed countries—especially in Africa. More recently, less-developed countries, anxious to develop their industry and to earn foreign exchange, have looked hopefully to the aluminum industry as one where they might, by taking advantage of local bauxite resources or favorable hydropower costs, export to industrial countries faced with constantly rising power costs. But instead of widening, as was then expected, this gap in power costs is more apt to be stabilized at so narrow a margin that trade barriers and investment disadvantages will preclude a large-scale shift to less-developed countries. Moreover, equalization of power costs among developed countries will permit location within such countries approximately in balance with demand. It might still be contended that comparative advantage should lead to some greater localization than this implies. Such a possibility cannot be precluded but it becomes impossible to analyze on a world basis and for a distant period.

Thus, at an international level, future location for the aluminum industry becomes very largely a matter of market orientation. This is not because markets exert a strong pull on the industry, but rather because ubiquitous moderate-cost power neutralizes the element which was expected to induce migration and because the trade and investment barriers faced by less-developed countries more than compensate for their small remaining advantage in real cost.

CHAPTER TWO

TREND OF ALUMINUM CONSUMPTION AND PRODUCTION

Long-run expansion of primary aluminum production will depend upon the growth of consumption. Future location of primary aluminum production has many determinants. One of the most important is the geographical distribution of markets which will influence production locations both because of the desire to reduce transportation costs and for reasons of strategy on the part of companies and governments in consuming countries. Therefore, it is necessary to examine both the expected over-all size of consumption and its expected geographical distribution before considering the prospective amount and location of new primary production capacity.

The present chapter will trace globally and by region the historical trends in consumption and production and will also look at trends in some of the industrial uses of aluminum. The following chapter will discuss future consumption under certain assumptions.

For a definition of consumption employed herein, the reader is referred to the appendix that follows this chapter.

TREND IN WORLD CONSUMPTION OF CRUDE ALUMINUM

Since the first utilization of the Hall–Héroult process for making aluminum in 1886, consumption has maintained a very rapid pace of growth. This can be seen from figures for the consumption of crude aluminum.[1] Crude aluminum consumption, estimated at 7,000 metric tons in 1900, advanced to a total of 122,000 metric tons by 1920. Since then growth has been much faster in absolute amounts,

---

[1] Crude aluminum as used in this section is almost entirely primary; secondary metal is included only to the extent that it constitutes a portion of an individual country's net imports in crude form. (See appendix to this chapter.)

reaching 823,000 metric tons by 1940 and 5,997,400 tons by 1964 (Table 1).

Table 1. Consumption of Crude Aluminum, Total and by Region, 1900-64   (thousand metric tons)

Year	World Total	Communist Bloc[a]	Rest of World					
			Total	Americas	West Europe	Asia	Africa	Oceania
1900	7.3	—	7.3	2.9	4.4	—	—	—
1901	7.5	—	7.5	3.2	4.3	—	—	—
1902	7.8	—	7.8	3.5	4.3	—	—	—
1903	8.2	—	8.2	3.4	4.8	—	—	—
1904	9.3	—	9.3	3.9	5.4	—	—	—
1905	11.5	—	11.5	4.3	7.2	—	—	—
1906	14.5	—	14.5	5.6	8.9	—	—	—
1907	14.8	—	14.8	5.0	9.8	—	—	—
1908	17.0	—	17.0	5.0	12.0	—	—	—
1909	35.3	—	35.3	15.5	19.8	—	—	—
1910	44.2	—	44.2	21.7	22.5	—	—	—
1911	46.8	—	46.8	20.9	25.9	—	—	—
1912	62.9	—	62.9	29.8	32.1	1.0	—	—
1913	66.1	—	66.1	31.2	34.6	0.3	—	—
1914	69.1	—	69.1	21.4	47.2	0.5	—	—
1915	63.2	—	63.2	28.9	33.8	0.5	—	—
1916	100.0	—	100.0	41.5	57.7	0.8	—	—
1917	128.5	—	128.5	42.4	85.1	1.0	—	—
1918	151.0	—	151.0	43.0	107.1	0.9	—	—
1919	132.6	—	132.6	61.5	69.6	1.5	—	—
1920	121.9	—	121.9	66.5	53.3	2.0	—	0.1
1921	71.3	—	71.3	37.5	31.7	2.0	—	0.1
1922	100.1	—	100.1	53.6	42.3	4.0	—	0.2
1923	140.3	—	140.3	75.6	60.4	4.0	—	0.3
1924	170.9	—	170.9	90.0	75.7	4.5	—	0.7
1925	175.9	—	175.9	80.0	90.6	5.0	—	0.3
1926	186.5	—	186.5	100.0	78.1	8.0	—	0.4
1927	199.9	—	199.9	100.0	93.6	6.0	—	0.3
1928	238.0	—	238.0	124.0	103.6	10.0	—	0.4
1929	269.6	8.7	260.9	137.0	110.2	13.0	—	0.7
1930	205.9	11.9	194.0	95.0	87.7	11.0	—	0.3
1931	173.7	21.5	152.2	70.0	77.1	5.0	—	0.1
1932	138.8	13.0	125.8	53.0	66.6	6.0	—	0.2

16

Table 1 (continued)

Year	World Total	Communist Bloc[a]	Rest of World					
			Total	Americas	West Europe	Asia	Africa	Oceania
1933	154.2	16.5	137.7	55.1	77.7	4.7	—	0.2
1934	223.8	22.5	201.3	80.7	113.3	6.8	0.1	0.4
1935	301.9	28.0	273.9	93.9	166.1	13.5	0.1	0.3
1936	397.5	43.5	354.0	135.4	199.4	18.5	0.1	0.6
1937	499.6	54.3	445.3	162.9	258.4	23.5	0.1	0.4
1938	505.4	68.0	437.4	88.5	298.2	51.7	0.2	0.3
1939	677.4	64.3	613.1	163.7	387.5	61.3	0.2	1.0
1940	822.8	65.5	757.3	224.4	481.4	51.0	0.1	1.2
1941	984.7	82.1	902.6	294.6	526.5	80.2	0.1	1.2
1942	1,365.8	109.0	1,256.8	566.9	575.6	110.2	0.1	4.0
1943	1,682.3	123.2	1,559.1	834.8	569.1	151.5	0.1	3.7
1944	1,468.2	222.1	1,246.1	646.1	473.8	122.2	0.1	4.0
1945	995.3	102.4	892.9	674.1	191.7	25.6	0.1	1.4
1946	945.5	107.1	838.4	573.8	248.0	14.3	0.4	1.9
1947	1,096.3	109.8	986.5	623.0	347.3	10.7	1.0	4.5
1948	1,249.1	133.2	1,115.9	731.0	365.2	13.7	1.0	5.0
1949	1,210.2	151.8	1,058.4	663.5	366.8	19.5	1.0	7.6
1950	1,583.6	244.3	1,339.3	904.0	399.4	27.6	2.0	6.3
1951	1,809.6	256.3	1,553.3	977.5	520.1	41.4	2.5	11.8
1952	1,957.3	273.0	1,684.3	1,060.6	571.1	41.1	3.0	8.5
1953	2,389.6	330.0	2,059.6	1,479.0	520.6	48.4	3.5	8.1
1954	2,542.5	444.0	2,098.5	1,370.2	651.5	60.6	5.0	11.2
1955	3,104.6	500.0	2,604.6	1,695.0	815.5	70.6	6.5	17.0
1956	3,227.3	526.0	2,701.3	1,735.7	855.7	83.8	6.1	20.0
1957	2,985.9	611.6	2,374.3	1,429.5	819.4	98.0	8.5	18.9
1958	3,170.6	690.2	2,480.4	1,483.3	851.8	109.3	9.2	26.8
1959	4,039.6	830.5	3,209.1	1,977.8	1,037.7	151.4	10.3	31.9
1960	4,178.9	940.0	3,247.8	1,715.0	1,281.0	198.4	14.0	39.4
1961	4,524.8	1,057.0	3,467.8	1,999.6	1,180.6	238.6	17.1	31.9
1962	4,979.1	1,091.4	3,887.7	2,301.1	1,256.0	258.2	19.0	53.4
1963	5,460.9	1,145.3	4,315.6	2,565.7	1,367.7	308.3	19.4	54.5
1964	5,997.4	1,220.0	4,777.4	2,794.9	1,533.1	348.4	23.0	78.0

[a] Most of these figures are estimates by Metallgesellschaft.
Note: Dashes in all cases indicate "no measurable amount."
Source: Metallgesellschaft Aktiengesellschaft, *Metal Statistics* (Frankfurt am Main, annual) and internal past records of the company.

In terms of growth rates, of course, the most rapid advances occurred during the infancy of the industry. Since 1920, however, the industry has been remarkable for its sustained high rate of growth. A United Nations study shows that the growth trend for crude aluminum lies between 10.4 per cent and 11.3 per cent per year, whether taken on a 10-, 30-, or 40-year basis terminating in 1959. Only the 20-year trend encompassing World War II yields a significantly different (lower) result.[2]

Growth in the non-Communist world has been at a slower pace, however, as the share of world consumption represented by the Communist bloc has increased. Thus, Communist-bloc countries accounted for under 10 per cent of the world total in 1939, but by 1964 they represented 20 per cent of the total. During the 1950–59 period non-Communist-bloc consumption of crude aluminum expanded at an annual rate of 8.8 per cent (and 8.1 per cent if consumption of all secondary metal is included).[3]

On a historical basis and still considering only consumption of crude aluminum, consumption has been dominated by North America and Western Europe. The two areas advanced about apace up to 1930, accounting for nearly all of the world total until the 1920's and about 90 per cent by 1930. During the 1930's war preparations in Europe, Russia, and the Japanese empire stimulated demand in those areas: by 1940 America took only a bit over one-fourth of all consumption and Western Europe more than twice this share, with Asia and the Soviet sphere countries using most of the rest. Just as the pre-World War II pattern was distorted by preparations for war, the postwar pattern was warped by its after effects. As of 1950, consumption in the Americas (including very small South American consumption) constituted 57 per cent of the world total and far surpassed Western Europe. This area had shrunk to about 25 per cent and the Communist bloc had grown to over 15 per cent. Growth during the 1950's brought some erosion of the American predominance and three major consuming areas emerged. Still, as of 1964, the American share was about 47 per cent of the world total, while Western Europe claimed 26 per cent

[2] United Nations ECOSOC, *World Economic Trends*, E/3629/E/CN.13/49 (New York, May 23, 1962), p. 24.
[3] *Ibid.*, p. 30. Such growth rates imply a doubling of consumption every 8 or 9 years.

of the metal and the Communist bloc about 20 per cent. Asia, with 7 per cent, consumed most of the remainder.

If these figures on consumption of crude aluminum are carried to the country level, the predominance of a few countries is immediately apparent. Outside of Europe and North America only Japan, India, Oceania, and possibly China, took as much as 1 per cent of the world total in 1964 (Table 2). Consumption in the United States was 42.4 per cent of the world total, while Canada added another 2.7 per cent. Of the Communist-bloc total of 1,220,-000 metric tons, Russia took 850,000 metric tons, 14.2 per cent of the world figure, with China, East Germany, Poland, Czechoslovakia, and Hungary each in the 1–2 per cent category. In Western Europe the highest consumption figure was reported by Germany at 6.5 per cent of the world total, followed by the United Kingdom at 6.0 per cent, France 4.2 per cent, Italy 2.0 per cent, and Belgium 1.9 per cent. Outside of Europe, Japan with 4.4 per cent of the total was the remaining major market. Lesser amounts were consumed in India, Australia, and Brazil, while Sweden, Switzerland, Austria, Yugoslavia, and Spain were other significant consumers of crude aluminum.

TRENDS IN THE CONSUMPTION OF ALUMINUM IN ALL FORMS
(COMMUNIST BLOC EXCLUDED)

The foregoing figures are based on world consumption of crude aluminum including only very little secondary metal. They provide a good approximation to the total world consumption of primary metal and to the industrial absorption of such metal by individual countries. For years since 1948 it is convenient to supplement these data with more complete figures which include secondary metal, thereby enlarging both world and individual country totals. For our final purposes there is less interest in a global figure which includes secondary. However, for individual countries, the total consumption, whether primary, secondary, or as finished products, is of direct interest; while world consumption of new primary metal will determine the amount of new capacity needed, the location of such capacity may be affected by the distribution of total demand for metal.

19

Table 2. Consumption of Crude Aluminum, Selected Countries, 1931-64

(thousand metric tons)

Year	United States	Canada	Brazil	Mexico	Other America	United Kingdom	Germany[a]	France	Italy	Belgium-Luxembourg	Sweden	Switzerland	Yugoslavia	Austria	Spain	Japan	India	Other Asia[b]	Africa	Oceania
1931	65.0	4.0	—	—	1.0	20.0	21.2	18.0	7.4	1.3	1.9	5.0	—	0.5	—	4.3	—	0.7	—	0.1
1932	48.0	4.0	—	—	1.0	17.5	18.1	15.0	5.5	1.0	1.4	4.5	—	0.4	—	5.3	—	0.7	—	0.2
1933	50.0	4.0	—	—	1.1	19.0	26.0	14.0	7.3	1.1	1.6	5.5	—	0.3	—	4.0	—	0.7	0.1	0.2
1934	74.0	5.5	—	—	1.2	23.0	48.8	18.0	9.4	1.1	2.4	6.5	—	0.4	—	5.8	—	1.0	0.1	0.4
1935	87.6	5.2	—	—	1.1	28.4	83.6	24.0	14.0	1.3	3.5	7.0	—	0.4	—	12.5	—	1.0	0.1	0.3
1936	127.0	7.0	—	—	1.4	35.0	102.3	27.0	17.0	2.4	2.4	9.0	—	0.5	—	17.0	—	1.5	0.1	0.6
1937	154.0	7.5	—	—	1.4	49.0	128.6	28.0	26.0	3.1	4.7	13.0	—	0.9	—	22.0	—	1.5	0.1	0.4
1938	81.2	6.0	—	—	1.3	45.0	171.5	31.3	25.6	2.0	6.5	11.5	—	1.0	0.7	50.0	—	1.7	0.2	0.3
1939	152.1	9.6	—	—	2.0	79.0	200.9	42.8	32.9	2.5	5.0	16.8	—	1.0	1.1	60.0	—	1.3	0.2	1.0
1940	205.9	16.5	—	—	2.0	103.6	240.6	67.9	42.8	1.0	5.0	14.2	—	1.0	1.4	50.0	0.1	0.9	0.1	1.2
1941	274.7	17.9	—	—	2.0	118.7	277.2	44.8	59.1	0.7	2.7	17.2	—	1.0	1.5	80.0	0.1	0.1	0.1	1.2
1942	535.2	29.7	—	—	2.0	198.4	254.3	39.2	56.6	3.3	2.6	14.4	—	1.0	0.8	110.0	0.1	0.1	0.1	4.0
1943	795.9	36.4	—	—	2.5	211.5	247.8	32.0	52.6	1.3	6.5	10.1	—	1.0	1.5	150.0	1.3	0.2	0.1	3.7
1944	608.8	34.8	—	—	2.5	152.5	266.3	25.1	8.5	1.0	4.8	8.2	—	1.0	1.6	120.0	2.0	0.2	0.1	4.0
1945	632.1	37.0	—	—	5.0	96.4	20.0	37.9	4.3	1.8	5.5	12.6	—	1.0	2.6	20.0	5.4	0.2	0.1	1.4
1946	535.4	30.7	3.0	0.2	4.5	116.4	10.0	56.6	10.3	2.5	11.6	20.7	—	2.0	3.1	3.0	10.8	4.0	0.4	1.9
1947	564.3	45.6	4.2	1.9	7.0	161.1	15.0	78.9	29.8	5.0	13.0	24.7	—	2.5	3.8	3.0	6.7	3.5	1.0	4.5
1948	659.0	59.4	5.6	1.0	6.0	176.2	24.0	86.3	23.8	3.6	13.3	15.2	—	3.5	2.6	7.0	4.2	5.5	1.0	5.0
1949	595.8	53.3	6.5	2.4	5.5	181.3	33.1	50.6	26.6	3.9	16.1	23.6	—	5.3	4.0	14.0	3.5	3.5	1.0	7.6

Year																				
1950	823.0	59.1	6.8	2.6	12.5	183.8	51.8	44.6	47.5	5.4	14.4	12.1	—	6.0	3.9	19.0	5.6	5.0	2.0	6.3
1951	877.1	78.2	11.3	3.4	7.5	206.2	81.5	79.0	47.3	14.0	18.2	29.9	2.8	12.0	5.7	30.8	7.1	3.5	2.5	11.8
1952	966.2	81.9	7.0	2.0	3.5	223.9	92.1	90.3	52.3	12.7	22.5	28.4	2.9	16.0	5.0	33.0	5.1	3.0	3.0	8.5
1953	1,377.6	80.3	8.6	5.0	7.5	183.5	102.1	74.9	48.2	12.2	18.8	23.9	2.8	23.6	7.1	38.5	4.4	5.5	3.5	8.1
1954	1,260.0	72.9	16.7	4.6	16.0	228.8	133.3	98.6	60.9	16.7	21.9	27.1	4.1	26.0	6.9	46.2	7.9	6.5	5.0	11.2
1955	1,581.8	83.0	8.2	8.0	14.0	290.0	176.7	108.5	61.9	26.2	30.2	35.4	8.9	33.5	12.7	49.6	13.0	8.0	6.5	17.0
1956	1,609.4	83.3	19.0	8.0	16.0	280.6	174.3	134.8	71.7	32.5	27.3	36.8	13.1	36.0	16.4	65.7	9.6	8.5	6.1	20.0
1957	1,313.4	70.7	21.6	8.8	15.0	216.8	185.7	153.1	66.2	27.2	25.4	39.3	17.1	32.7	21.0	74.7	13.8	9.5	8.5	18.9
1958	1,334.0	92.4	26.0	12.4	18.5	236.2	191.1	143.1	62.7	41.1	32.4	31.8	16.2	31.4	27.1	80.7	15.6	13.0	9.2	26.8
1959	1,845.3	80.5	27.0	9.5	15.5	293.6	231.8	167.8	83.0	48.5	36.0	42.2	25.1	34.4	37.1	112.6	22.9	15.0	10.3	31.9
1960	1,541.2	109.6	33.0	11.2	20.0	359.6	313.0	212.7	99.0	63.5	37.9	47.9	40.9	37.0	20.0	150.5	24.9	23.0	14.0	39.4
1961	1,791.5	123.2	36.8	10.5	38.0	284.1	297.0	202.2	107.5	69.0	33.6	47.5	30.9	37.6	20.0	185.6	30.0	23.0	17.1	31.9
1962	2,089.3	127.7	41.2	15.9	27.0	286.7	301.9	235.6	115.0	67.5	39.7	53.5	38.7	38.0	22.5	184.2	49.0	25.0	19.0	53.4
1963	2,340.3	142.2	46.7	14.5	22.0	318.6	315.4	242.5	128.0	88.3	50.7	48.3	41.7	39.0	35.2	219.3	62.0	27.0	19.4	54.5
1964	2,534.9	160.0	50.0	18.0	32.0	358.9	385.9	249.3	120.0	111.5	51.2	53.8	45.0	43.0	46.0	261.8	56.6	30.0	23.0[c]	78.0

[a] German figure is for the Federal Republic after 1945.

[b] Excludes Communist-bloc countries.

[c] As of 1964, this was divided 16.0 South Africa and 7.0 rest of Africa.

Source: Metallgesellschaft Aktiengesellschaft, *Metal Statistics* (Frankfurt am Main, annual), and internal past records of the company.

The consumption figures by country which follow extend to all identifiable aluminum, including secondary metal and fabricated products entering trade.[4] On this basis the consumption of certain countries, especially in Europe, is significantly altered in two ways. The inclusion of secondary metal enhances the position of several major industrial countries, while more inclusive trade data often tend to diminish the position of other such countries. In addition, certain nonindustrial countries whose consumption of crude aluminum is small or nil are more important consumers when all aluminum is included.

Of these two influences, the inclusion of secondary metal is of greatest significance. Secondary metal production, which serves as an approximation to its consumption, was 19.5 per cent of total estimated non-Communist consumption in 1964. It had been as high as 26.1 per cent in 1948 but fell in the 18–20 per cent range after 1954. Its distribution is quite unequal, however, since in many countries there is no measurable production of secondary metal while in others it is a very significant share of the total. Thus in 1964, 34.6 per cent of German consumption was secondary, 33.2 per cent in the United Kingdom, 32.6 per cent in Italy, and 28.5 per cent in Japan. For the United States the ratio was only 16.3 per cent in that year.[5]

Inclusion of data on net trade in all identifiable aluminum products tends to raise apparent consumption in less-industrialized countries and also in certain industrialized countries which lack a

[4] Chiefly United Nations Standard International Trade Classification numbers 284 and 684, but also relevant categories of 691, 692, 693, 697, and 698 where shown. While figures on this basis provide a closer approximation to the domestic demand for metal, a country may still be a net importer or exporter of final products containing aluminum but not identified as such in trading statistics. To that extent, its domestic demand for metal in end uses would be greater or lesser than our figures show. Since industrial specialization in the production of end products containing aluminum will persist irrespective of the geographical pattern of demand for such products, the figures employed may be more useful as a guide to the industrial demand pattern than pure domestic end-use consumption, including all trade in products containing aluminum, would be.

[5] Figures based on Metallgesellschaft Aktiengesellschaft, *Metal Statistics* (Frankfurt am Main, 1964), p. 7, and on Table 4 of this chapter. One reason for the low ratio of secondary consumption in the United States is that primary smelters have preempted a portion of the market for metal used in castings for the automotive industry by supplying liquid metal direct to the foundries at prices competitive with secondary.

primary aluminum industry. This reflects the tendency for less-industrialized countries to import metal in fabricated form because they often lack a sufficiently large market to justify a full range of facilities for forming crude metal. Somewhat similar considerations may apply to smaller industrial countries which lack a primary industry. Where access to these markets is relatively free, they may purchase semi-fabricated metal from primary producing countries.

Moreover, there is some tendency for fabrication to thrive in primary producing countries and to seek export outlets for finished products. Accordingly, the more inclusive trade data raise consumption totals in nearly all nonindustrial countries, e.g., India and Mexico, as well as in such places as Denmark, Finland, Sweden, and the Netherlands, while reducing the apparent figure in such exporters of fabricated products as Austria, France, Germany, Japan, Switzerland, and the United Kingdom. Belgium is most sharply affected; although it has no primary industry, it is a large-scale importer of metal in crude form which is processed into fabricated products for export. Therefore, Belgian apparent domestic consumption shrinks abruptly when adjustment is made for trade in finished products. The United States displays no clear pattern; prior to 1963 more inclusive net trade figures have generally provided a small increment to the consumption total, but in 1963 and 1964 the reverse was true.

While inclusion of secondary metal significantly affects the non-Communist-world total, the use of more comprehensive trade figures simply affects the distribution of this total.[6] Using the non-Communist-world consumption of crude metal plus secondary production as an approximation to total consumption on a continental basis, it is quickly apparent that while inclusion of secondary metal somewhat improves the share of consumption in Europe and Asia at the expense of the Americas, the broad regional pattern of distribution is not greatly affected. Thus, in 1964 non-Communist crude metal consumption was distributed as follows: Americas 58.5 per cent, Europe 32.1 per cent, Asia 7.3 per cent, Oceania 1.6

[6] A small amount of net exports of crude aluminum from the Communist bloc is included in non-Communist consumption. Any other Communist-bloc exports are certainly small. Thus, no special allowance must be made for the effect of Communist-bloc trade in finished products on the total.

per cent, and Africa 0.5 per cent. If secondary metal is included, the Americas' share falls to 55.7 per cent. Europe's share rises to 34.6 per cent, and Asia's to 7.8 per cent (Table 3).

Table 3. Approximate Non-Communist-World Total Aluminum Consumption, by Region, 1948, 1964

Region	Total[a] (thousand metric tons)		Per Cent of Total	
	1948	1964	1948	1964
World	1,510.8	5,933.2	100.0	100.0
Americas	997.4	3,307.4	66.0	55.7
Europe	490.6	2,052.0	32.5	34.6
Asia	16.5	464.8	0.9	7.8
Africa	1.0	26.0	0.1	0.4
Oceania	5.3	83.0	0.4	1.4

[a] Includes consumption of crude aluminum plus local secondary production. Sources: Crude consumption from Metallgesellschaft Aktiengesellschaft, *Metal Statistics* (Frankfurt am Main, annual). Secondary taken from United Nations, *Statistical Yearbook*, or Metallgesellschaft.

Further adjustment by continents to allow for more comprehensive trade data is difficult because of data gaps, but it is apparent that such adjustment would raise the American, African, and Oceanian shares at the expense of Europe—the adjustments in each case being less than one percentage point.

On a country basis, however, more significant shifts of the sort suggested above occur when allowance is made both for secondary metal and for more comprehensive trade data. Thus the major shifts in this distribution from that derived from crude metal consumption alone are the higher ranking of the United Kingdom, Japan, and Italy, while Belgium in particular is of diminished importance. The position of nonindustrial countries also would be greater than when only crude metal is considered, but data gaps preclude a complete analysis of this difference which in world terms is, in any case, not large. After all adjustments, non-Communist consumption in 1964 was distributed as follows among the chief consumers: United States 50.6 per cent, Germany 9.2 per cent, United Kingdom 8.7 per cent, Japan 6.2 per cent, France 4.1 per cent, Canada 3.2 per cent, Italy 3.0 per cent, Oceania 1.4 per cent, India 1.3 per cent, Sweden 1.1 per cent, Austria 1.0 per cent, Switzerland 1.0 per cent, Spain 0.9 per cent, Belgium 0.8 per cent,

Brazil 0.7 per cent, with the remaining 6.8 per cent dispersed among a number of countries. It is worth noting that outside the Communist bloc as of 1964, only three countries that appeared to have total consumption of 20,000 tons or more (Denmark, Belgium, and the Netherlands) (Table 4) still lacked primary production facilities.

The 1964 distribution of consumption reflects the effects of growth in total non-Communist consumption amounting to 126 per cent between 1954 and 1964 or a rate of 8.5 per cent per year. This is somewhat less than the 9.7 per cent rate of 1948–54. On a regional basis the growth rate for the 1954–64 period declined from the 1948–54 period for all areas except Oceania. In the Americas, dominated by the United States and Canada, the 1954–64 rate of growth was 7.3 per cent in contrast to the 8.7 per cent of the earlier period. However, the growth of European consumption also moderated during this period (from 10.5 per cent to 8.6 per cent) as did the small African consumption. In Asia, due largely to the Japanese performance, the annual growth rate remained at an astonishing 19.8 per cent per year for 1954–64, down from the 29.0 per cent rate of the earlier period.[7]

On a national basis during the 1954–64 period consumption has expanded most rapidly in Japan (21.0 per cent per year) among the major consumers. Among the nonindustrial countries a number have achieved very high growth rates, but the pattern is not uniform. In Europe the best growth rates were recorded by countries with comparatively low initial consumption levels such as Greece, Spain, Yugoslavia, Finland, and Denmark. Within Europe, e.g., there has been some tendency for per capita consumption rates to equalize (Table 4).

As between continents, the differential growth rates have produced an ambiguous result. Although the more rapid growth rate of low-consumption areas (Asia, Africa) than of high-consumption areas (Europe, Americas) has reduced the relative gap somewhat, it has not succeeded in diminishing the absolute differences because of the extremely low initial figures in areas of low consumption and the small absolute changes which have occurred in their consumption while much greater absolute growth was occurring in developed areas.

[7] These figures refer to crude consumption plus secondary production in each case.

Table 4. Total Consumption[a] and Per Capita Consumption of Aluminum, Selected Countries, 1948-64

(A=thousand metric tons; B=kg per capita)

Country	1948	1949	1950	1951	1952	1953	1954	1955	1956	1957	1958	1959	1960	1961	1962	1963	1964
United States																	
A	876.0	736.9	1,033.2	1,147.8	1,248.5	1,727.3	1,529.6	1,892.2	1,923.8	1,643.6	1,610.2	2,206.9	1,856.5	2,118.2	2,516.1	2,778.3	3,001.1
B	6.0	4.9	6.8	7.4	7.9	10.8	7.4	11.4	11.4	9.6	9.2	12.4	10.3	11.5	13.5	14.7	15.6
Canada																	
A	67.1	60.5	55.3	75.8	70.4	77.9	67.5	79.7	98.3	78.8	81.0	69.6	96.8	106.6	136.6	171.4	188.3
B	5.1	4.5	4.0	5.4	4.9	5.3	4.4	5.1	6.1	4.7	4.7	4.0	5.4	5.8	7.3	9.1	9.8
Brazil																	
A	8.3	10.3	10.1	15.4	12.2	13.8	19.3	8.7	20.5	22.6	26.8	27.7	31.6	39.1	41.5	47.2	44.1
B	0.2	0.2	0.2	0.3	0.2	0.2	0.3	0.1	0.3	0.4	0.4	0.4	0.4	0.5	0.6	0.6	0.6
Mexico																	
A	—	—	—	—	3.7	6.6	6.1	11.3	9.2	9.8	13.2	10.5	12.4	11.0	18.0	18.6	21.1
B	—	—	—	—	0.1	0.2	0.2	0.4	0.3	0.3	0.4	0.3	0.4	0.3	0.5	0.5	0.5
Venezuela																	
A	9.1	—	—	—	8.8	10.2	12.7	8.1	13.5	11.7	17.1	14.7	8.7	10.8	8.7	7.7	8.2
B	—	—	—	—	1.6	1.8	2.2	1.3	2.1	1.8	2.5	2.1	1.2	1.4	1.1	1.0	1.0
Colombia																	
A	—	—	—	2.4	1.2	2.5	—	3.9	6.1	3.7	3.7	4.3	4.6	4.6	5.8	7.8	7.5
B	—	—	—	0.2	0.1	0.2	—	0.3	2.1	0.3	0.3	0.3	0.3	0.3	0.4	0.5	0.5
Germany																	
A	24.7	75.1	97.9	116.2	122.5	138.9	187.5	249.1	240.1	254.2	269.5	315.6	419.2	404.7	412.3	437.8	545.7
B	0.5	1.6	2.0	2.4	2.5	2.8	3.8	5.0	4.7	4.9	5.2	6.0	7.9	7.5	7.5	7.9	9.7
United Kingdom																	
A	209.9	195.4	220.3	238.8	264.3	218.0	274.1	345.6	335.5	279.4	303.0	358.1	440.9	377.1	387.0	444.2	516.3
B	4.2	3.9	4.4	4.7	5.2	4.3	5.4	6.8	6.5	5.4	5.8	6.9	8.4	7.1	7.2	8.3	9.5

France																	
A	107.5	62.4	57.0	91.8	109.5	89.4	121.1	127.0	161.3	181.9	165.1	178.9	219.2	204.7	242.3	233.5	240.7
B	2.6	1.5	1.4	2.2	2.6	2.1	2.8	2.9	3.7	4.1	3.7	4.0	4.8	4.4	5.2	4.9	5.0
Italy																	
A	36.6	38.8	62.4	62.7	65.3	68.3	84.0	83.9	91.3	87.0	85.0	110.0	141.6	152.4	167.9	185.0	178.0
B	0.8	0.8	1.3	1.3	1.4	1.4	1.8	1.7	1.9	1.8	1.7	2.2	2.9	3.1	3.3	3.7	3.5
Switzerland																	
A	12.2	20.5	9.2	26.2	24.5	18.0	23.3	23.3	36.7	39.3	34.3	45.2	42.4	46.5	54.6	52.7	60.6
B	2.7	4.5	2.0	5.6	5.1	3.7	4.8	4.7	7.3	7.7	6.6	8.5	7.9	8.5	9.6	9.1	10.1
Sweden																	
A	18.5	19.5	15.6	22.3	27.6	22.9	29.6	37.4	34.9	32.5	39.3	42.2	47.7	42.2	47.6	57.4	63.3
B	2.7	2.8	2.2	3.1	3.9	3.2	4.1	5.1	4.8	4.4	5.3	5.7	6.4	5.6	6.3	7.6	8.2
Austria																	
A	4.1	6.2	7.1	10.0	13.3	26.0	28.7	38.9	39.4	41.7	42.1	41.6	45.4	49.7	52.3	53.8	60.8
B	0.6	0.9	1.0	1.4	1.9	3.7	4.1	5.6	5.6	6.0	6.0	5.9	6.4	7.0	7.4	7.6	8.6
Belgium-Luxembourg																	
A	—	—	—	12.7	11.0	11.1	14.1	16.8	21.3	14.1	22.1	24.8	33.2	32.5	28.8	39.1	49.6
B	—	—	—	1.4	1.2	1.2	1.5	1.8	2.3	1.5	2.4	2.6	3.5	3.4	3.0	4.2	5.3
Netherlands																	
A	—	—	—	—	11.8	14.1	15.6	17.6	20.2	17.6	15.7	23.7	28.7	25.1	26.4	35.4	38.7
B	—	—	—	—	1.1	1.3	1.5	1.6	1.9	1.6	1.4	2.1	2.5	2.2	2.2	3.0	3.2
Yugoslavia																	
A	—	—	—	3.5	5.0	3.1	4.9	6.7	11.0	14.0	4.9	21.3	42.5	30.5	37.6	35.5	43.8
B	—	—	—	0.2	0.3	0.2	0.3	0.4	0.6	0.8	0.3	1.2	2.3	1.6	2.0	1.9	2.3
Norway																	
A	8.0	9.1	10.2	11.9	13.2	14.3	14.4	15.3	15.4	16.7	14.9	15.7	22.3	23.4	25.0	26.3	34.7
B	2.5	2.8	3.1	3.6	4.0	4.2	4.2	4.5	4.4	4.8	4.3	4.4	6.2	6.5	6.9	7.1	9.4

Table 4 (continued)

Country	1948	1949	1950	1951	1952	1953	1954	1955	1956	1957	1958	1959	1960	1961	1962	1963	1964
Spain																		
A	2.6	4.0	3.9	5.7	5.0	7.6	8.6	13.2	16.7	21.2	27.6	27.6	31.9	17.4	21.9	30.3	54.0
B	0.1	0.1	0.1	0.2	0.2	0.3	0.3	0.5	0.6	0.7	0.9	0.9	1.1	0.6	0.7	1.0	1.7
Finland																		
A	—	—	—	—	4.1	4.4	5.0	9.5	9.6	9.9	10.8	17.0	17.0	16.7	17.6	14.0	18.5
B	—	—	—	—	1.0	1.1	1.2	2.3	2.2	2.3	2.4	3.9	3.9	3.7	3.9	3.1	4.0
Denmark																		
A	—	5.3	5.9	7.3	6.0	5.2	6.9	9.0	8.2	9.8	10.8	15.5	17.3	15.8	18.0	18.6	23.6
B	—	1.3	1.4	1.7	1.4	1.2	1.6	2.0	2.8	2.2	2.4	3.4	3.8	3.5	3.8	4.0	5.0
Japan																		
A	—	—	14.2	28.8	34.0	46.6	52.8	48.6	73.2	90.7	97.6	141.4	190.1	245.3	244.9	304.5	366.0
B	—	—	0.2	0.3	0.4	0.5	0.6	0.5	0.8	1.0	1.1	1.5	2.0	2.6	2.6	3.2	3.8
India																		
A	13.0	10.3	13.5	13.3	9.8	9.3	17.6	23.4	17.6	29.6	27.9	36.7	43.6	44.0	74.1	n.a.	78.2
B	b	b	b	b	b	b	0.1	0.1	b	0.1	0.1	0.1	0.1	0.1	0.2	n.a.	0.2
New Zealand																		
A	—	—	—	—	5.6	—	—	3.4	2.8	3.5	5.3	4.8	5.6	6.3	8.0	9.5	11.7
B	—	—	—	—	2.8	—	—	1.6	1.3	1.6	2.3	2.1	2.3	2.6	3.2	3.8	4.5

[a] Total consumption aims to encompass all identifiable aluminum. It has been arrived at by starting with the Metallgesellschaft figure for crude consumption, adding domestic secondary production, and adjusting for trade in fabricated products.
[b] Less than 0.05 kg.

Sources: Based on Metallgesellschaft Aktiengesellschaft, *Metal Statistics* (Frankfurt am Main, annual), and internal past records of the company; United Nations and U.S. Department of Commerce trade figures; and United Nations, *Demographic Yearbook*.

These trends may be seen also in per capita consumption figures. Despite a strong growth rate, per capita consumption in India was less than 0.2 kg per head in 1964, up from an insignificant figure for 1954. Meanwhile, Greece moved up from 0.2 kg to 1.0 kg, Yugoslavia from 0.3 kg to 2.3 kg, Finland from 1.2 kg to 4.0 kg, Spain from 0.3 kg to 1.7 kg, and Japan from 0.6 kg to 3.8 kg.

END USES OF ALUMINUM

It is worthwhile to look at trends in the end use of aluminum in different countries as a rough guide to the reasonableness of over-all projections arrived at by other methods.

Because it is an industrial material, aluminum finds its way into a broad spectrum of final products. However, it is conventional to classify these into a relatively small number of end-use categories that absorb most of the metal. This has been done by percentage in each category for the United States and most Western European countries as of 1964 in Table 5.[8] The countries represented account for over 85 per cent of all non-Communist consumption and therefore provide a fair approximation to the total world end-use pattern.

In the United States the two leading uses in 1964 were construction and transportation. Together they constituted over half of all domestic consumption. Electric industries and consumer durables followed in importance at about 10–12 per cent each, while packaging and machinery and equipment constituted smaller totals.

In Europe, for the countries shown, transportation was the leading domestic consumer, with about 30 per cent of the total. It was followed by the electrical industries and by machinery and equipment industries, each with 11–13 per cent of the total, and by construction and by packaging with about 10 per cent each.

Thus, the principal difference between the two distributions is the relatively greater importance of aluminum as a construction material in the United States at the expense of most other categories that rank higher in Europe. To the extent that the categories are comparable, it appears that European consumption in kg per capita does not touch the American level in any major use.[9] The

[8] Note the figures are for all aluminum, including secondary.

[9] In fact these data are assembled by industry trade associations and are quite variable as to quality, coverage, and definition. The Organisation for Economic Co-operation and Development (OECD) is now making an effort to make the figures more comparable and of uniform quality.

discrepancy is most marked in the case of construction, where the American consumption amounted to 4.4 kg per capita compared with only 0.6 kg in Europe.

Table 5. Distribution of Aluminum Consumption by Domestic End Use, Selected Countries, 1964

Country	Per Cent						
	Building Products	Transportation	Electrical	Consumer Durables	Containers and Packaging	Machinery and Equipment	Other
United States	24.9	25.3	12.1	9.6	8.7	8.0	11.4
Major European:	10.2	29.9	13.3	8.4	9.5	11.1	17.5
Germany	9.2	29.2	17.1	3.6	9.3	14.9	16.8
United Kingdom	8.9	32.0	13.1	10.8	7.0	9.0	19.2
France	8.2	32.3	12.4	11.5	10.1	10.2	15.2
Italy	12.5	42.5	6.3	8.1	9.4	9.6	11.7
Austria	10.4	7.3	15.6	4.5	21.4	6.5	34.3
Netherlands	18.4	10.7	11.6	15.5	17.5	13.0	13.2
Belgium	19.0	7.7	3.1	8.0	12.9	10.1	39.2
Norway	25.3	10.8	17.0	17.3	9.0	3.2	17.3
Japan	10.5	22.2	6.5	28.3	2.4	11.8	18.4

Source: Organisation for Economic Co-operation and Development, *The Non-Ferrous Metals Industry, 1964* (Paris, 1965), pp. 68-69.

Within Europe there are wide differences in the way aluminum is employed. To a certain extent this reflects the industrial pattern of the countries concerned. Thus, Italy devoted 42.5 per cent of its metal in 1964 to the transportation equipment category and France, the United Kingdom, and Germany also registered high in this use (all have important automobile industries). Countries without these industries used their aluminum elsewhere: in electrical industries (Norway), in construction (Norway, Belgium, and Holland), in packaging (Austria and Holland), or in consumer appliances (Norway and Holland).

More detailed breakdowns are available for some countries. The predominance of passenger cars as consumers of aluminum in France is evident from the fact that in 1961 they took 22 per cent

30

of all consumption or about 70 per cent of that going into the transport category. Approximately two-thirds of the electrical engineering category in France was for wires, cables, and bus bars and the rest was mostly power station equipment. In most other categories consumption was dispersed among various subclasses of use. In Austria, electrical transmission equipment, roofing, siding, and door and window frames were principal end uses.[10]

In the United States figures have only recently been published beyond the level of detail shown in Table 5. Windows, doors, and screens are the principal items in the building products group, followed by siding and roofing. Heating and ductwork take a growing share. In the transportation category, passenger cars are the largest category. Aircraft and missiles and commercial vehicles follow in order. Consumption in marine and rail equipment is still quite small. The consumer durable category is widely dispersed over appliances, utensils, and furniture, with refrigeration and air conditioning being the principal users. The big electrical user is transmission equipment, which takes well over two-thirds of the category. In other categories consumption is well dispersed.[11]

It is useful to contrast the consumption pattern in the United States and Europe with that of certain Asian countries. Japan, as the largest consumer in Asia, displays some of the characteristics of Western industrial countries in its diversity of use. In 1964 the largest use was for consumer durables which accounted for 28.3 per cent of the total. This was followed by transportation at 22.2 per cent, with remaining consumption well dispersed. The principal difference from Europe is the greater stress on consumer durables.

The electrical industries may constitute a significant share of consumption in lesser-developed countries. Thus, in India as of 1960, it is estimated that over one-third of all consumption was in such industries, followed by household and "commercial" supplies and transportation.[12] India is engaged in a continuing program of electrification and, for the purpose of conserving foreign exchange, has sought to substitute aluminum for copper in electrical uses. In countries where this receives less emphasis and where the general

[10] Organisation for Economic Co-operation and Development, *Non-Ferrous Metals, 1961-1962* (Paris, September 1962), pp. 134-45.
[11] The Aluminum Association press release, March 5, 1965.
[12] Indian figures from United Nations ECAFE, *Bauxite and Mineral Resources of Asia and the Far East,* Mineral Resources Development Series No. 17 (Bangkok, 1962).

level of industry is less advanced, the use of aluminum in the electrical industries may be negligible. The use of aluminum in construction, especially as roofing sheet, occupies a prominent place in less-developed tropical countries. In other instances luxury apartments and prestige office buildings may be relatively large consumers of aluminum in an otherwise undeveloped market. This is oftentimes the case in Latin America, where a Western-oriented upper class provides a market for such buildings, or in some other areas which have become regional centers for European or American investments. Lacking major automotive or aircraft industries, underdeveloped countries rarely make significant use of aluminum in transportation. In some cases, specialized agricultural or fishing products for export may result in a relatively high share of consumption in containers in some low-income countries. More often than not, however, low-income countries, unless engaged in electrification, may appropriately be characterized as "a pots and pans market"—a term used to describe essentially undeveloped, low-tonnage markets for aluminum.

On a historical basis it is difficult to trace the shifting pattern of end uses because of the lack of consistent historical series. In a context of general growth in consumption, it is apparent that the use of aluminum has been broadened greatly from a fairly narrow range of early specialty uses. Wallace has traced the early history of use of the metal in the United States and noted the early predominance of use as a deoxidant in steel making, for pots and pans, and for electrical transmission.[13] More significantly, however, he attributes much of its rapid growth to the simultaneous expansion of the auto industry before and after World War I which took about half of all prewar shipments and as high as 70 per cent in 1922.[14] In pre-World War I years, utensils made up about one-third of the market, with transmission cable and deoxidant (along with previously mentioned autos) constituting the rest. Post-World War I saw the auto share surge still further but decline thereafter as sheet steel and cast iron replaced aluminum.[15] However, new uses

[13] Donald H. Wallace, *Market Control in the Aluminum Industry* (Cambridge: Harvard University Press, 1937), pp. 14-23 and 61-68. In 1903 about 37 per cent of all sales was wire (p. 16).

[14] *Ibid.*, p. 19. Indeed the consumption per new car in that year apparently was higher than the level 30 years later—a fact which should give some pause to enthusiastic trend projectors!

[15] *Ibid.*, p. 62.

in aircraft, consumer durables and construction helped to maintain the industry's growth. As of 1930, transportation took 38 per cent of the total and electrical equipment and utensils each 16 per cent.[16] It is surmised that the depression brought decreased consumption in autos and relatively more in the electrical industry. By 1938 transportation remained the leading category at 29 per cent, followed by machinery at 15 per cent and utensils at 14 per cent.[17] Throughout postwar years building has been the most significant end use of aluminum, except for the brief flurry of heavy military use during the Korean War. Transportation, which had fallen to only about 10 per cent of the total in 1948, has returned to its position of importance.[18]

Early consumption trends in Europe were somewhat similar to the United States with kitchen equipment the largest consumer in pre-World War I Germany, and with autos (the chief post-World War I consumer) said to account for 75 per cent of use in England, France, and Switzerland in 1921.[19] The 1930's saw the rapid build-up of the German industry and certainly much of this went for military purposes, as was no doubt the case in other European countries such as France and Italy. By 1953 the characteristic postwar pattern had been fairly well established in Europe with transportation taking 23.6 per cent of consumption followed by electricity at 14.5 per cent and packaging with 11 per cent. Use in the building industry was only 6.7 per cent,[20] a share which has grown somewhat since.

TREND IN PRIMARY PRODUCTION

In broad outline the location of production parallels that of consumption which already has been discussed. However, it is

---

[16] *Ibid.*, p. 64.

[17] Cited in Merton J. Peck, *Competition in the Aluminum Industry* (Cambridge: Harvard University Press, 1961). N. H. Engle, H. E. Gregory, and R. Mossé, *Aluminum* (Chicago: Richard D. Irwin, 1945), p. 252, have a slightly different distribution for the 1935-39 average. However, transportation is still the leading category at 22.2 per cent.

[18] Cited in Dominick and Dominick, *An Analysis of the Aluminum Industry of North America* (New York, 1962), Table VII.

[19] Wallace, *op. cit.*, pp. 65-66.

[20] Organisation for European Economic Co-operation, *The Non-Ferrous Metal Industry in Europe* (Paris, 1955), p. 79.

worthwhile to look briefly at the trend and current pattern of production of the primary industry.[21] In addition, it will be helpful as background for subsequent discussion to examine the trend and existing location of production of bauxite and alumina.

Virtually all primary metal was produced in Europe and North America until the late 1930's when small amounts of production appeared in the Japanese empire and the U.S.S.R. The industry was of comparable magnitude in the two main centers until the 1930's. At that time the severity of the depression in the United States and the beginning of war preparations in Europe caused European production to move well ahead. As of 1939, world production of primary was distributed as follows: Europe 54.5 per cent, North America 32.5 per cent, Japan 4.3 per cent, and U.S.S.R. 8.6 per cent (Table 6). During the years of World War II, North American production quickly surpassed that of all of the rest of the world; in 1943 at 1,284,500 metric tons, it amounted to 66 per cent of the world total. In immediate postwar years output fell in all areas, but by 1950 was everywhere on the ascent. In that year North American output remained high at 67.1 per cent of the world total, while Europe produced 16.3 per cent, the Communist bloc 14.5 per cent, and Asia 2.0 per cent. In the following decade the American share (including minor production in Brazil) fell and was 51.3 per cent as of 1964, while Western Europe rose to 19.8 per cent, the Communist bloc to 21.1 per cent, Asia to 5.6 per cent, and Africa and Australia appeared with minor production of 0.9 per cent and 1.3 per cent respectively. If only the non-Communist total is considered, the share of the Americas was 65.1 per cent, Western Europe 25.1 per cent, and Asia 7.1 per cent.[22]

In Europe, France and Switzerland were the early important producers. The United Kingdom industry also began early but was stunted at the level of primary production by the absence of extensive low-cost power and by the British trade policy. Germany and Norway assumed important roles after World War I. Except for the United Kingdom, all of these countries have remained im-

---

[21] Secondary production is significant only in countries which are major consumers of metal, since the chief source of scrap is that generated in various stages of fabrication. Likewise, such countries will provide the greatest supplies of obsolete metal.

[22] Derived from Table 6.

Table 6. Production of Primary Aluminum, Total and by Region,
1900-64   (thousand metric tons)

| Year | World Total | Com-mu-nist bloc[a] | Rest of World | | | | | |
			Total	Amer-icas	West Europe	Asia	Af-rica	Oce-ania
1900	7.3	—	7.3	3.2	4.1	—	—	—
1901	7.5	—	7.5	3.2	4.3	—	—	—
1902	7.8	—	7.8	3.3	4.5	—	—	—
1903	8.2	—	8.2	3.4	4.8	—	—	—
1904	9.3	—	9.3	3.9	5.4	—	—	—
1905	11.5	—	11.5	4.5	7.0	—	—	—
1906	14.5	—	14.5	6.0	8.5	—	—	—
1907	19.8	—	19.8	8.0	11.8	—	—	—
1908	18.6	—	18.6	6.0	12.6	—	—	—
1909	31.2	—	31.2	16.0	15.2	—	—	—
1910	43.8	—	43.8	19.6	24.2	—	—	—
1911	45.0	—	45.0	20.3	24.7	—	—	—
1912	62.6	—	62.6	27.8	34.8	—	—	—
1913	66.2	—	66.2	27.3	38.9	—	—	—
1914	69.0	—	69.0	33.1	35.9	—	—	—
1915	77.9	—	77.9	49.6	28.3	—	—	—
1916	104.2	—	104.2	60.7	43.5	—	—	—
1917	124.2	—	124.2	70.7	53.5	—	—	—
1918	134.5	—	134.5	71.6	62.9	—	—	—
1919	132.7	—	132.7	73.3	59.4	—	—	—
1920	126.2	—	126.2	74.6	51.6	—	—	—
1921	77.8	—	77.8	32.5	45.3	—	—	—
1922	92.0	—	92.0	43.6	48.4	—	—	—
1923	138.0	—	138.0	68.5	69.5	—	—	—
1924	168.7	—	168.7	80.8	87.9	—	—	—
1925	181.4	—	181.4	77.1	104.3	—	—	—
1926	195.2	—	195.2	83.6	111.6	—	—	—
1927	220.4	—	220.4	112.7	107.7	—	—	—
1928	256.1	—	256.1	135.5	120.6	—	—	—
1929	281.7	—	281.7	145.4	136.3	—	—	—
1930	268.9	—	268.9	138.8	130.1	—	—	—
1931	219.3	0.1	219.2	111.5	107.7	—	—	—
1932	153.6	0.9	152.7	65.5	87.2	—	—	—
1933	141.6	4.4	137.2	54.5	82.7	—	—	—
1934	170.9	14.4	156.5	49.2	106.6	0.7	—	—

Table 6 (continued)

Year	World Total	Communist bloc[a]	Rest of World Total	Americas	West Europe	Asia	Africa	Oceania
1935	258.3	24.8	233.5	74.7	154.1	4.7	—	—
1936	365.5	38.7	326.8	128.2	191.1	7.5	—	—
1937	492.7	48.8	443.9	175.0	258.4	10.5	—	—
1938	589.3	52.5	536.8	194.7	319.7	22.4	—	—
1939	686.7	59.3	627.4	223.5	374.5	29.4	—	—
1940	783.2	69.6	713.6	286.1	392.2	35.3	—	—
1941	1,036.6	82.5	954.1	474.4	416.8	62.9	—	—
1942	1,393.8	72.8	1,321.0	781.7	450.4	88.9	—	—
1943	1,949.1	92.9	1,856.2	1,284.5	447.9	123.8	—	—
1944	1,709.7	107.7	1,602.0	1,123.6	357.9	120.5	—	—
1945	869.9	93.2	776.7	645.1	112.2	19.4	—	—
1946	774.3	93.0	681.3	547.2	127.6	6.5	—	—
1947	1,051.1	96.5	954.6	790.0	158.6	6.0	—	—
1948	1,225.3	110.5	1,114.8	898.6	203.3	12.9	—	—
1949	1,257.1	130.7	1,126.4	882.7	217.7	26.0	—	—
1950	1,506.9	219.0	1,287.9	1,011.9	245.7	30.3	—	—
1951	1,807.6	237.1	1,570.5	1,165.2	361.6	43.7	—	—
1952	2,032.3	259.8	1,772.5	1,304.8	417.5	50.2	—	—
1953	2,453.9	318.3	2,135.6	1,632.1	449.3	54.2	—	—
1954	2,820.4	427.3	2,393.1	1,835.3	492.6	65.2	—	—
1955	3,104.7	514.2	2,590.5	1,973.4	543.8	72.0	—	1.3
1956	3,342.6	548.0	2,794.6	2,092.1	611.8	81.4	—	9.3
1957	3,397.2	645.1	2,752.1	2,009.1	640.5	84.1	7.6	10.8
1958	3,554.7	740.5	2,814.2	2,011.2	658.4	101.6	31.9	11.1
1959	4,092.0	853.9	3,238.1	2,334.5	724.7	125.0	42.3	11.6
1960	4,547.9	935.6	3,612.3	2,537.0	859.9	159.7	43.9	11.8
1961	4,588.3	1,068.7	3,519.6	2,347.1	930.7	181.1	47.6	13.4
1962	4,990.7	1,128.8	3,861.9	2,569.3	1,006.4	217.8	52.2	16.5
1963	5,429.2	1,182.1	4,247.1	2,776.9	1,085.7	289.7	52.9	41.9
1964	6,085.7	1,284.7	4,801.0	3,124.6	1,203.2	341.4	51.5	80.3

[a] Most of these figures are estimates by Metallgesellschaft. The U.S.S.R. and some other Communist-bloc countries do not publish aluminum production figures.

Note: Dashes in all cases indicate "none."

Source: Metallgesellschaft Aktiengesellschaft, *Metal Statistics* (Frankfurt am Main, annual), and internal past records of the company.

portant producers and have been joined by others as well. In 1964 France produced 26.3 per cent and Germany 18.3 per cent of the West European total. Meanwhile, significant production appeared in Italy by about 1930, and Austria, Spain, and Yugoslavia have important production dating from war and post-World War II years. Norway, not a great consumer of metal, is a major producer. Its rapidly growing production surpassed the German figure in 1962, and announced expansions will make Norway the largest West European producer in the future (Table 7).

In the Americas, the United States has led the way with three-fourths of the total, but Canadian production has been a significant factor. The Canadians produce nearly one-fourth of the North American total, most of it for export. The only other production in the Americas is small amounts in Brazil and recently in Mexico and Surinam, although a smelter is planned for Venezuela and possibly Curaçao. Elsewhere outside the Communist bloc, Japan is the largest producer. Growing industries are to be found in Australia and India, and Taiwan continues with a small output. Cameroon completes the list of those non-Communist countries now producing primary aluminum metal.

The addition of secondary metal raises the production figures of some of the European countries in particular. Total production is shown for selected countries in Table 8.

Perhaps the most interesting thing about the distribution of production is its rough tendency toward conformity with consumption if viewed on a continental basis. As of 1964 the Americas consumed 47 per cent of the world's primary metal used and produced about 51 per cent. Europe had the opposite pattern, consuming 26 per cent and producing 20 per cent, while the Communist-bloc position was almost exactly balanced. Regionally at least, the chief centers of aluminum consumption so far have been able to produce the metal.

On a country-by-country basis there is, of course, greater departure from this rule since the United Kingdom, Germany, and Belgium are significant consumers who must import a substantial portion or all of their consumption while Canada, Norway, and Cameroon are significant producers but much smaller consumers. The Canadian–United Kingdom relationship reflects the historical

Table 7. Production of Primary Aluminum by Country (Excluding Communist Bloc), 1900, 1910, 1920, 1930-64

(thousand metric tons)

Year	United States	Canada	Mexico	Brazil	France	Norway	Germany[a]	Italy	Austria	Switzerland[b]	Spain	Yugoslavia	United Kingdom	Sweden	Japan	India	Taiwan	Cameroon	Australia
1900	3.2	—	—	—	1.0	—	0.7	—	—	1.8	—	—	0.6	—	—	—	—	—	—
1910	16.1	3.5	—	—	9.5	0.9	1.1	0.8	—	6.9	—	—	5.0	—	—	—	—	—	—
1920	62.6	12.0	—	—	12.3	5.6	10.5	1.2	—	14.0	—	—	8.0	—	—	—	—	—	—
1930	103.9	34.9	—	—	26.0	27.4	30.1	8.0	3.0	20.5	1.1	—	14.0	—	—	—	—	—	—
1931	80.5	31.0	—	—	18.0	21.4	27.1	11.1	2.5	12.2	1.2	—	14.2	—	—	—	—	—	—
1932	47.6	17.9	—	—	14.5	18.0	19.3	13.4	2.1	8.5	1.1	—	10.3	—	—	—	—	—	—
1933	38.6	15.9	—	—	14.5	15.5	18.9	12.1	2.1	7.5	1.1	—	11.0	—	—	—	—	—	—
1934	33.6	15.6	—	—	16.2	15.5	37.2	12.8	2.2	8.2	1.2	—	13.0	0.3	0.7	—	—	—	—
1935	54.1	20.6	—	—	21.9	15.3	70.8	13.8	2.5	11.6	1.3	—	15.1	1.8	4.7	—	—	—	—
1936	102.0	26.2	—	—	26.5	15.4	97.5	15.9	3.3	13.6	0.7	—	16.4	1.8	7.5	—	—	—	—
1937	132.8	42.2	—	—	34.5	23.0	127.6	22.9	4.4	25.0	—	0.2	19.0	1.8	10.5	—	—	—	—
1938	130.1	64.6	—	—	45.3	29.0	161.4	25.8	4.4	26.5	0.7	1.3	23.4	1.9	17.8	—	4.6	—	—
1939	148.4	75.1	—	—	52.7	31.1	195.2	33.9	4.3	27.0	1.1	1.8	25.4	2.0	21.7	—	7.7	—	—
1940	187.1	99.0	—	—	61.7	27.8	204.9	38.6	6.7	28.3	1.3	2.0	19.3	1.6	26.5	—	8.8	—	—
1941	280.4	194.0	—	—	63.9	17.5	212.4	48.8	21.4	25.6	1.1	2.0	23.0	1.1	50.4	—	12.5	—	—
1942	472.7	309.0	—	—	45.2	20.5	227.3	45.4	36.8	23.7	0.7	2.0	47.5	1.3	75.4	—	13.5	—	—
1943	834.8	449.7	—	—	46.5	23.5	204.2	46.2	46.0	18.5	0.8	2.0	56.6	3.6	108.0	1.3	14.5	—	—
1944	704.4	419.2	—	—	26.2	20.0	192.7	16.8	51.6	9.7	0.2	1.0	36.0	3.7	109.5	1.8	9.2	—	—
1945	449.1	195.7	—	0.3	37.2	4.6	19.7	4.3	5.2	5.0	0.6	—	32.4	3.2	16.5	2.3	0.6	—	—

Year																			
1946	371.6	175.4	—	0.2	47.8	16.7	0.7	11.0	1.0	13.1	1.0	0.6	32.1	3.6	3.2	3.3	—	—	—
1947	518.7	271.3	—	—	53.4	21.7	0.8	25.1	4.5	18.5	1.0	1.3	29.4	2.9	2.7	3.3	—	—	—
1948	565.6	333.0	—	—	63.7	31.0	7.5	32.6	13.3	19.0	0.5	1.9	30.5	3.3	7.0	3.4	2.5	—	—
1949	547.5	335.2	—	—	53.9	34.6	29.1	25.9	14.8	21.0	1.2	2.5	30.8	3.9	21.2	3.5	1.3	—	—
1950	651.9	360.0	—	0.4	60.6	45.3	27.8	37.0	18.0	19.0	2.2	1.9	29.9	4.0	24.8	3.7	1.8	—	—
1951	759.2	405.6	—	1.1	91.1	50.3	74.1	50.8	26.4	27.0	4.2	2.8	28.2	6.7	36.9	3.9	2.9	—	—
1952	850.3	453.4	—	1.2	106.1	51.1	100.5	52.8	36.7	26.9	4.1	2.6	28.5	8.2	42.7	3.6	3.9	—	—
1953	1,135.8	495.1	—	1.5	113.0	53.2	106.9	55.5	43.5	29.0	4.4	2.8	31.4	9.6	45.5	3.8	4.9	—	—
1954	1,325.0	508.8	—	1.7	120.2	61.3	129.2	57.6	48.0	26.0	4.1	3.5	32.1	10.6	53.1	5.0	7.1	—	—
1955	1,420.4	551.3	—	6.3	129.0	72.1	137.1	61.5	57.2	30.2	10.4	11.5	24.8	10.0	57.5	7.3	7.2	—	1.3
1956	1,523.1	562.7	—	8.9	149.8	92.7	147.4	63.7	59.4	30.1	13.5	14.7	28.0	12.5	66.0	6.6	8.8	—	9.3
1957	1,494.8	505.4	—	11.9	159.9	95.6	153.8	66.2	56.4	31.1	15.9	18.1	29.9	13.6	68.0	7.9	8.2	7.6	10.8
1958	1,420.2	579.1	—	18.1	169.1	121.6	136.8	64.1	56.9	31.5	16.2	21.7	26.8	13.7	84.6	8.4	8.6	31.9	11.1
1959	1,772.7	543.7	—		173.0	144.9	151.2	75.0	65.6	34.3	21.1	19.2	24.9	15.5	100.1	17.4	7.5	42.3	11.6
1960	1,827.5	691.3	—	18.2	235.2	165.4	168.9	83.6	68.0	39.0	29.3	25.1	29.4	16.0	133.2	18.2	8.3	43.9	11.8
1961	1,727.0	601.6	—	18.5	279.2	171.9	172.6	83.4	67.7	42.2	37.6	27.4	32.8	15.9	153.7	18.4	9.0	47.6	13.4
1962	1,921.4	626.2	—	21.7	294.5	205.9	177.8	80.9	74.1	49.6	44.7	28.0	34.6	16.3	171.5	35.2	11.1	52.2	16.5
1963	2,097.9	652.6	5.5	20.9	298.4	219.2	208.8	91.4	76.5	60.1	46.1	35.9	31.1	18.2	223.9	53.9	11.9	52.9	41.9
1964	2,316.0	764.8	17.8	26.0	316.0	261.5	219.9	115.0	77.7	64.2	49.6	34.8	32.2	32.3	265.8	56.2	19.4	51.5	80.3

[a] German figure is for the Federal Republic after 1945.
[b] Includes Austria until 1920.
Source: Metallgesellschaft Aktiengesellschaft, *Metal Statistics* (Frankfurt am Main, annual), and internal past records of the company.

Table 8. Total Production of Aluminum (Including Secondary), Selected Countries, 1948-64

(thousand metric tons)

Country	1948	1949	1950	1951	1952	1953	1954	1955	1956	1957	1958	1959	1960	1961	1962	1963	1964
United States	825.8	711.5	872.9	1,024.6	1,126.6	1,470.2	1,589.9	1,725.2	1,831.3	1,823.0	1,682.9	2,099.2	2,126.4	2,035.7	2,340.3	2,556.6	2,804.5
Canada	337.2	339.0	365.0	413.0	460.9	502.1	514.0	558.2	569.2	512.3	585.8	550.7	699.6	608.8	633.9	661.9	783.8
Germany	51.8	72.0	83.3	127.7	149.3	155.5	199.2	229.8	234.6	243.5	238.6	264.3	302.6	308.0	320.5	365.2	408.5
France	88.9	74.5	84.4	112.2	126.7	136.2	147.1	160.6	182.1	196.5	209.7	217.5	279.2	322.8	341.3	348.0	366.3
Norway	31.0	35.7	47.0	52.1	52.8	56.2	64.1	74.6	95.2	98.2	124.3	148.9	168.3	174.9	209.0	221.9	270.0
United Kingdom	112.0	107.2	111.0	100.7	102.8	113.2	117.5	121.9	125.7	128.6	127.4	133.9	140.8	151.8	166.4	180.1	203.8
Italy	45.8	38.9	49.2	67.5	69.0	75.5	81.0	84.5	85.7	89.2	90.6	105.0	125.6	128.4	136.9	156.4	173.0
Austria	13.3	14.8	18.0	26.4	36.7	50.0	57.2	67.8	70.8	72.9	74.3	85.2	90.1	92.9	102.5	108.1	113.1
Switzerland	21.5	23.5	21.5	30.0	29.9	32.0	29.0	34.2	35.6	38.1	37.5	40.3	45.6	49.2	58.1	71.1	78.7
Sweden	3.8	4.5	4.6	7.3	9.0	10.9	12.4	13.1	14.4	15.7	16.6	16.3	16.7	16.7	17.3	19.7	33.8
Japan	9.8	23.1	28.4	41.7	50.8	56.1	65.5	67.1	80.5	87.4	108.1	137.1	182.7	223.2	243.0	314.7	370.2
Australia	0.3	0.3	0.4	0.4	0.4	0.4	0.4	1.7	9.8	11.2	14.4	15.8	16.3	18.2	21.5	46.9	85.3

Source: United Nations, *Statistical Yearbook*, and Metallgesellschaft Aktiengesellschaft, *Metal Statistics* (Frankfurt am Main, annual).

effect of the Commonwealth tie and British trade policy in this area. The effect of energy-cost should not be entirely discounted, however. Each of the major importing countries is without abundant cheap hydro or gas energy while each of the major exporters has built its industry on this basis. There is in this partial confirmation of the kind of locational influence which one might expect, but the more striking feature is the degree to which countries or regions so far have met their needs through domestic or regional production.

While production of primary metal tends to conform to the location of markets, the raw material stages of the industry have become more remote from markets. At the alumina stage the growing divorce between locus of production and consumption is a fairly recent development. At the bauxite stage the separation reflects the increased demand for the raw material and the inadequacy or absence of bauxite resources in metal-consuming countries. A brief look at these two stages and their significance for location is in order.

TRENDS IN BAUXITE PRODUCTION

In 1964 about 19 per cent of the world bauxite production was attributed to the Communist bloc while the remainder of some 26.4 million metric tons was produced in other countries. Of the non-Communist-world total, the principal source was the islands of the Caribbean Sea and the bordering coast of South America. Together they produced 57.5 per cent of this total. Jamaica led with 29.6 per cent, followed by Surinam 15.1 per cent, and Guyana 8.1 per cent, with lesser amounts from the Dominican Republic and Haiti. Western Europe produced another 19.1 per cent of the non-Communist total with France providing 9.2 per cent, Yugoslavia 4.9 per cent, and Greece 4.0 per cent. Africa provided 6.8 per cent, mostly from Guinea (5.4 per cent). The Asian total was slightly lower (6.5 per cent) and well dispersed among Malaya, India, and Indonesia. The United States, by far the largest consumer of bauxite, produced only 6.0 per cent of the non-Communist-world total in 1964.[23] Australia, still not a large producer with only 3.4 per cent in 1964, will expand rapidly to become a major supplier in the next few years.

[23] Foregoing figures from Metallgesellschaft Aktiengesellschaft, *op. cit.* (1965), p. 4.

The current pattern of bauxite production differs greatly from the earlier pattern. In earlier years of the industry industrial countries supplied most of the bauxite. At the end of the 1920's, with a primary industry about equal to America's, Europe supplied 59 per cent of the world bauxite total and the United States nearly 20 per cent; most of the remainder was divided between British Guiana (now Guyana) and Dutch Guiana, but largely oriented to North American smelters. Of the European figure, France produced more than one-half and the remainder came from Italy, Hungary, and Yugoslavia. By 1937, in line with its faster growth in primary metal, a still higher percentage of the enlarged output came from Europe with 67.4 per cent of the total; the French share was now only one-third of the European total with Hungary, U.S.S.R., Yugoslavia, and Italy following in order and Greece now a significant factor. Elsewhere the shares of the United States and the Guianas fell off (an absolute decline in the case of British Guiana) and significant new production occurred in the Dutch East Indies.[24]

By 1948 West European production still lagged behind the 1937 level, although for Europe as a whole output was above prewar. The more significant change was in the relative positions of two major South American producers who supplied the now dominant North American smelters. Guyana and Surinam, which constituted 12.2 per cent of the world total in 1937, both expanded their output strongly during the World War II and immediate postwar years so that by 1948 they produced 49 per cent of the world total, or 55.3 per cent of the non-Communist output, while the West European share of this latter total fell to only 15.7 per cent, being surpassed by the United States with 20.2 per cent.[25]

Following 1948 the most significant development was the appearance of Jamaica as a sizeable producer in 1952–53 and its quick ascent to the number one position, supplying the voracious North American smelters. Elsewhere, Surinam and Guyana have maintained or expanded their large output, and Guinea (particularly after 1960), Greece, Yugoslavia, France, Malaya, and the Dominican Republic have expanded strongly. U.S. production displays no strong trend in either direction but it has steadily diminished in

---

[24] Mineraux et Metaux, *Renseignments Statistiques, 1938* (Paris, 1938), pp. 89-90.

[25] U.S. Bureau of Mines, *Minerals Yearbook, 1951* (Washington: U.S. Government Printing Office, 1954), p. 206.

terms of world importance and in its ability to supply American smelters.[26]

Thus to summarize on recent trends: the emergence of tropical producers to their positions of transcendent importance in World War II and postwar years and the imminent appearance of Australia as a major producer have brought about a revolution in the sources of supply of bauxite. This reflects the inability of industrial countries, particularly North America, to meet expanding needs for ore. *Prima facie* it might be expected that this shift would exert some influence on the location of subsequent stages of the industry. So far the effect is most noticeable at the alumina stage.

LOCATIONAL PATTERN OF ALUMINA PRODUCTION

The location of alumina production again parallels the location of primary metal on a regional basis. Of approximately 7,200,000 metric tons of alumina produced in 1960 in non-Communist countries almost half was produced in the United States (48.2 per cent) with another 14 per cent in Canada. Still within the North American ambit but outside the industrial countries was the 9.3 per cent in Jamaica, much of it in support of the Canadian industry. The remaining 28.5 per cent in the non-Communist world was mostly in Europe (19.8 per cent) where France (8.2 per cent) and Germany (6.0 per cent) were predominant. Japan produced 4.9 per cent of the total and the rest was dispersed, with minor production in India. Thus, on a continental basis the production of alumina corresponds quite closely to primary metal output. There are several notable exceptions at the country level, however. In 1960 Norway, for example, had 4.6 per cent of the non-Communist metal total but had no significant alumina production. Jamaica and Guinea, with 9.3 per cent and 2.5 per cent of alumina output respectively, did not produce metal. Canada produced 19.1 per cent of the metal

[26] Metallgesellschaft Aktiengesellschaft, *op. cit.* (1965), p. 4. It is interesting to note that a rough kind of regional balance still exists between supply and consumption of bauxite, for Western Europe with 25.1 per cent of non-Communist primary metal production in 1964 also provided 19.1 per cent of the bauxite and had access to African supplies, while the Americas produced 65.1 per cent of the metal (nearly all in North America) and 64.3 per cent of the bauxite (mostly in the Caribbean area) and the Asian figures were also in approximate balance.

but only 14.0 per cent of the alumina. The United States also had a slightly higher share of metal output at 50.6 per cent than of alumina production, 48.2 per cent.[27]

The above pattern of alumina production is now in a state of transition, however. Output in Guinea has been expanded since 1960 and major new plants are planned or being built in Australia, Surinam, Virgin Islands, and Greece. A projected Rotterdam plant apparently has been abandoned in favor of increased output in Surinam. These projects illustrate the current trend of alumina production to follow the bauxite source so as to save on transportation. This was done first by Alcan for its Jamaican operation. An international consortium with Olin-Mathieson as the principal holder acquired a similar operation in Guinea. Alcoa will shortly complete a facility in Surinam; an international consortium headed by Kaiser is building one in Australia where Alcoa already has a moderate-sized plant and Alusuisse seems likely to follow, and a consortium headed by Pechiney will build one in Greece as part of an integrated industry there.[28] These moves respond to the shifting geography of bauxite production. They are undertaken cautiously in response to real economic incentives but sometimes in the face of less welcome political exposure.

THE INTERNATIONALIZATION OF PRODUCTION

The existing pattern of the aluminum industry then is one of heavy concentration of production as well as consumption in industrialized countries. This pattern has prevailed from the very beginning of the industry. In recent years, however, the production pattern has begun to change, most markedly in the bauxite and alumina phases of the industry. At the bauxite and alumina stages the changes reflect a less advantageous resource position in industrialized countries in the face of greatly increased demand and a desire to minimize transport costs. As yet, however, it is uncertain to what extent the changing resource pattern will affect the location of reduction.

[27] Alumina figures from United Kingdom Overseas Geological Surveys, *Bauxite, Alumina and Aluminium* (London: H.M.S.O., 1962), p. 47. For 1962 list of capacity, see *Metal Bulletin*, "Aluminium World Survey" Special Issue (London, December 1963), pp. 127-29. Aluminum production from Metallgesellschaft Aktiengesellschaft, *op. cit.* (1964), p. 5.

[28] *Metal Bulletin*, "Aluminium World Survey," *op. cit.*, pp. 127-29, and subsequent press reports.

As we have seen, from the early days of the industry up to World War II the industry at all stages was concentrated in industrial countries, with even most of the bauxite coming from such countries. However, as the industry grew prior to World War II it sought bauxite further afield from its original centers in the United States and Western Europe, expanding first in other European countries and beginning the exploitation of the reserves of the Guianas. This reflected the increased demand for bauxite and the strategy of producing companies to secure long-term supplies under their own corporate control.

Within the chief producing countries of Europe and North America it has up to now proved feasible to provide sufficient moderate or low-cost electric power to permit the expansion of reduction facilities. In the United Kingdom this was not the case and early reliance was made on imports of metal from Canada and Norway to supply an extensive aluminum fabricating industry.

Further expansion of aluminum production in post-World War II years brought increased dependence on distant bauxite sources, and the locational implications of this were accepted by the industry, resulting in several lines of action. The Canadians, without domestic bauxite and in the process of expanding reduction facilities on the West Coast distant from their Quebec alumina plant, were the first to build alumina facilities next to foreign bauxite sources in Jamaica. American producers, also growing dependent on Caribbean sources, initially were content to locate their alumina plants at Gulf Coast ports as an intermediate step, but more recently some have built closer to ore sites. In Western Europe the expanding European output of bauxite met most needs. Norway, lacking either bauxite or alumina plants, has relied upon imported alumina. By the late 1950's European companies were showing increased interest in foreign bauxite sources and in advantageously located alumina plants for processing such ore. Thus, European firms invested in Guinea and explored possibilities in Australia, Greece, and the Guianas in order to secure their ore supplies. A major alumina plant was erected in Guinea, primarily serving European markets.

In addition to changing resource patterns, conditions of metal supply and considerations of corporate strategy have affected the location of primary production in a minor way in recent years. During the early 1950's aluminum was in short supply both in Europe and North America and there was no great push for foreign markets. The appearance of excess capacity late in the decade, combined with diminished restrictions on trade, encouraged a quest for foreign markets. Insofar as the motive was only to ensure an outlet for their metal, major integrated firms were content to form subsidiaries or seek mergers with existing fabricators, both domestically and abroad. However, the resulting exposure has quickened the interest both of governments and firms in production in foreign markets. Countries able to offer a market large enough to absorb the output of a small reduction plant or able to offer advantageous conditions for an export industry have sought mills— examples are Mexico, Surinam, Greece, Cameroon, and Argentina. Meanwhile subsidiaries or jointly owned reduction facilities were established by the major international firms in such industrial countries as Japan, Australia, and Norway.

Thus, changing resource positions and altered market conditions combined with the aspirations of nonindustrial countries have caused the aluminum industry to become more international in scope. While consumption remains heavily concentrated in the industrial countries, the location of production is becoming a more open question. Undoubtedly further expansion in industrial countries will remain the most important factor in the growth of the industry, but for the first time significant expansion in nonindustrial countries must be considered a realistic possibility because of the changing resource pattern and continued growth in size of the market.

APPENDIX: DEFINITION OF CONSUMPTION EMPLOYED

There are problems of a conceptual and a statistical nature inherent in almost any definition of consumption. These problems are especially acute in the case of metals, where losses occur as the material passes through several stages of refining, forming, and fabrication, is incorporated in other products (often of a durable nature), and may finally return in the form of scrap for reuse. Except in the case of destructive uses, it is not possible to select some point in this process and unequivocally say the meal has been "consumed."[29]

[29] For discussion of these issues, see H. H. Landsberg, L. L. Fischman, and J. L. Fisher, *Resources in America's Future* (Baltimore: Johns Hopkins Press for Resources for the Future, Inc., 1963).

The choice of a definition of consumption will depend upon the purpose for which it is intended, allowing always for the convenience of the choice as affected by data availability. In the present case the purpose is to estimate future demand for primary aluminum so as to arrive at the amount of primary capacity to be built over the period and to estimate future demand by area for aluminum in all forms to assist in determining future production locations. Fairly rough figures will suffice, in contrast with the more refined estimates that might be called for when considering short-term production decisions.

For the first purpose (i.e., amount of new plant capacity), the figure sought is simply the global demand for primary aluminum which is required to produce the end products containing aluminum, net of temporary inventory and stockpile changes and after allowing for processing losses and for the fact that some amount of secondary metal will be available. For individual countries where decisions concerning production location may be at stake we cannot be content with such a figure. The interest here is rather in the market for all of the primary, secondary, and fabricated metals required to meet local end product needs, without regard to type or source.

Unhappily the future world consumption of primary aluminum cannot be projected in a single step because in most applications the demand is not for primary but rather for raw aluminum, irrespective of whether it is of primary or secondary origin. So long as secondary metal is satisfactory for the purpose and available at a competitive price, it will inhibit expansion in primary consumption.[30] Therefore, the future estimates must consider total aluminum consumption, and a further step is required to distinguish the portion of this which will be supplied by secondary metal, the demand for primary being treated as a residual.

At what point in the stream should metal be thought of as consumed? From the standpoint of the producers (the ones who must make investment decisions), when the metal (whether primary or secondary) is acquired by industrial users the producer has realized a genuine sale. This is also a statistically convenient point at

---

[30] Note that secondary metal may be a somewhat inferior substitute for primary metal. However, in some applications (chiefly castings) it is wholly adequate and, so long as supply does not exceed demand in those applications to which it is suited. secondary may be considered to substitute pound for pound for primary.

which to measure consumption. This concept of "industrial absorption" excludes from consumption shipments to the producer's own plants and net additions to the producer's inventory or to government stockpiles.[31] Included in consumption, however, are losses and scrap generated at the various stages of fabrication after the metal has left the producer. Some of this metal may appear for reprocessing (along with old scrap) but then is viewed as newly consumed when it passes through the cycle again.

Such a concept is satisfactory for the world as a whole; exclusive of estimated secondary metal it also yields a good measure of the world demand for primary. However, where the focus is on an individual country and where trade therefore enters the picture, the concept is less satisfactory. If consumption is measured at the point of acquisition by industrial users, a country which is a net exporter of fabricated metal or of products containing aluminum appears as a large consumer of aluminum while the net importer of such products might not appear at all as a consumer. This result may be acceptable as a description of existing industrial production patterns, but if market figures are to be used to shed light on location decisions they should not ignore such net trade patterns.

To make matters more difficult, as the process of fabrication proceeds, aluminum often loses its identity and becomes impossible to trace statistically. There is no very good solution to this problem, but it seems best to pursue consumption by country as far as the figures allow.

Therefore, total consumption figures for individual countries employed here have been based on the following concept: Primary production, less net inventory increase (to the extent available), less net increase in stockpiles, plus secondary production, plus net imports of all identifiable aluminum equals total apparent consumption. For the world as a whole, consumption consists of primary production less increase in stock (private or public) plus production of secondary metal.

Because past data are not available for all countries it is difficult to apply these concepts to all areas. For continents, however, spot

---

[31] The term is from United Nations ECOSOC, *op. cit.*, p. 15. Note that some of the series employed in that study exclude most secondary metal from the consumption totals, but the concept can be applied as desired either to primary or to total metal consumed.

checks suggest that the Metallgesellschaft figure for consumption of crude metal plus production of secondary metal yields a total very similar to that achieved after more elaborate attention to trade in fabricated metal. In practice, this series on aluminum consumption maintained by Metallgesellschaft has been employed as the basis for historical analysis by continent, with certain modifications in recent years (Table 3).[32] More complete data for trade would lower the European consumption figure slightly and raise that of other areas.

The Metallgesellschaft figure for individual countries in effect is consumption of primary aluminum (Table 2). Since it includes in consumption only that part of secondary metal which enters international trade in crude form (a small figure), it offers a fair approximation to the absorption of primary metal. This series is the best available for an examination of the more remote historical trend.

When summed, this figure is comprehensive enough to use for the consumption of primary metal in the world as a whole, but it is a less satisfactory measure of consumption for individual countries, principally because it neglects most secondary metal and makes no attempt to trace the destination of fabricated products. Therefore, in arriving at consumption figures by country for more recent years, whenever possible we have added secondary production to the Metallgesellschaft figures by country and have substituted more inclusive trade data from the United Nations and from other sources for the Metallgesellschaft figure on net imports (Table 4). For individual countries the concept then differs from Metallgesellschaft because of the inclusion of domestically produced secondary metal in consumption and the inclusion of all identifiable net aluminum imports instead of merely imports of crude metal. The discrepancies between the two sets of figures for individual countries on account of the different definitions of imports are significant for certain countries in Europe in particular. For broader geographical areas, after allowing for the inclusion of secondary metal, there is little difference between the two figures.[33]

[32] See Metallgesellschaft Aktiengesellschaft, *Metal Statistics* (Frankfurt am Main, annual).
[33] Cf., p. 38 ff.

49

CHAPTER THREE

FUTURE DEMAND FOR ALUMINUM

The foregoing has shown the trend of aluminum consumption and production. It is essential as background to an understanding of future patterns of consumption. Nevertheless, however well grounded in the past, any venture into the unknown is precisely that.

Statistical techniques for estimating future consumption may vary from simple trend projection to elaborate multiple correlation procedures, but in each case they rely upon the perseverance of past relationships. Selection among possible techniques involves an implicit judgment concerning the plausibility of this perseverance. On the other hand, complete abandonment of statistical methods of projection would result in a highly personal figure that would not be accorded credibility. Over the years industry and other sources have had a very spotty record of success in making estimates of future consumption by a variety of statistical techniques.

There is no reason to suppose that we can improve on this record, and fortunately, so far as this study is concerned, no definitive forecast of future demand for aluminum is required. More sophisticated forecasts of demand would be justified if a long lead time were required in order for supply to react to increases in demand. For example, if it were concluded that low-cost hydroelectricity would be a decisive influence on the industry's costs and location, it would then be necessary to devote much attention to demand because the selection and development of hydro sites is a slow and costly process and short-term imbalances of supply and demand would be hard to correct. Since in our judgment the industry will make increasing use of conventional or nuclear thermal power which demands a much shorter lead time, the accuracy of long-term forecasts becomes of less importance.

51

Our major conclusion about location will be compatible with any probable range of demand. Before settling on a figure to be used in subsequent analysis it may be useful to discuss some of the influences which might be expected to affect demand and the resulting problems for any would-be forecaster.

CONSIDERATIONS IN FORECASTING ALUMINUM DEMAND

Aluminum is an industrial material possessing certain physical characteristics that make it suitable for use in end products. However, in few instances are its characteristics so unique or the design requirements of end products so rigid as to require aluminum. Therefore, aluminum must face the competition of other materials. But neither aluminum nor competing materials are immutable—new alloys, new fabricating processes, and other technological changes can alter the very physical properties of the materials themselves, as well as their relative costs. Neither are the end product markets unchanging; shifts in national incomes, in industrial production and in the product mix of final goods will affect the demand for aluminum and for competing materials. These relationships, and especially future alterations in them, are too complex to be dealt with confidently by any forecasting technique. However, a qualitative discussion of their effects may be helpful in providing perspective to figures derived by statistical techniques.

If the only change that occurs in this set of interacting relationships is an increase in national income we might expect aluminum consumption to increase in proportion to the income gain. In fact, however, a change in national income induces simultaneous changes in the relative importance of manufacturing and in the mix of final products and services demanded. Some of these changes would be likely to occur in the absence of income change, but they may be speeded and accentuated by growth of income.

The effect on demand of a shifting product mix is much more difficult to discern; it could act either to accelerate or dampen demand, depending upon whether the shift is toward products which employ aluminum. Moreover, the direction of changes in product mix is inherently difficult to foresee. At one stage of development it is likely that income growth will induce a shift in the pattern of consumption away from food and clothing and in favor of industrial goods containing aluminum. Hence, aluminum consumption might grow faster than income. At a more advanced stage the shift may prove to be in the direction of consumption of services, and the changes in consumption pattern induced by income changes would not allow aluminum consumption to keep pace with income growth. Not every country need follow this pattern, however. Apart from income changes, consumer tastes alter in an unpredictable fashion and the effect of such changes on the demand for products containing aluminum cannot be forecast.

Still, assuming that we know how incomes will change and that we know the distribution of final demand that will result at the changed income level, we face yet another problem. What materials will be employed in what proportions to meet this demand pattern? Here we face the problem of substitution—the fact that alternative materials will do approximately the same job. If their physical properties were unchanging and their relative prices likewise fixed, then it should be possible to indicate in what proportions they would be used once these relationships were fully comprehended. In fact, however, materials are constantly being modified or new ones developed to compete with existing materials and changes in the technology of their production or the institutional environment in which they are produced and sold affect their relative prices. Therefore, the competitive relationships between materials also change in a fashion that is difficult to predict. There may be some hope of forecasting future changes in income and some prospect of discerning prospective cost relationships, but on a global basis it is almost impossible to foresee changes in the product mix or in the technological possibilities of utilizing various materials in production.

Discomfort with aggregate statistical techniques has led to efforts to break demand down into its various end use components so as to consider the forces bearing on each.[1] It would be a forbidding task to attempt analysis of prospective end use in each market of each country and to sum them for an aggregate figure of future demand. To be wholly convincing, such analysis would re-

[1] This practice is quite common among the aluminum companies analyzing various national markets. See also N. H. Engle, H. E. Gregory, and R. Mossé, *Aluminum* (Chicago: Richard D. Irwin, 1945), and James E. Rosenzweig, *The Demand for Aluminum: A Case Study in Long-Range Forecasting*, University of Illinois Bulletin (Urbana, 1957).

quire detailed estimates of future income and consumption patterns and knowledge of local conditions of supply of competitive materials. This would be a most difficult assignment even if done for one country, for it implies the construction of a matrix encompassing other materials as well. Attempts to short-cut this procedure have not been notably successful. However, the previously recounted pattern of end-use trends (Chapter 2, pp. 29–33) does afford some useful insight into the plausibility of the estimates otherwise derived.

Since the building up of estimates of the end use of aluminum must be rejected here for the reasons just mentioned, it is necessary to fall back onto the two other methods of making aggregate estimates: trend projection and correlation techniques.

An excellent study of prospective aluminum demand making use of these techniques is contained in the United Nations Economic and Social Council publication, *World Economic Trends, A Study of Prospective Production and Demand for Primary Commodities.*[2] Given the limitations of data availability, the enormous diversity in conditions from one country to another, and the inherent uncertainty of any forecasting technique, it is doubtful that more elaborate techniques would improve on the aggregate results derived in that study. It will be useful, however, to discuss their derivation so as to suggest how different assumptions might affect the figures. Also, the United Nations study provides only global figures which must be allocated to smaller areas by some procedure.

The United Nations forecasts are based on relationships prevailing during 1950–59. For the period 1959–64, total non-Communist consumption of primary increased at about 8.3 per cent per year or 8.5 per cent if secondary is included. This is slightly above the United Nations trend. The most recent years include a sustained period of rapid economic growth in the United States and may overstate the longer range expectations.

Irrespective of the method used, the non-Communist world growth rate for consumption falls in the 6–8 per cent range with most of the estimates concentrating in the middle or higher portion of the range.

[2] United Nations ECOSOC, E/3629/E/CN.13/49 (New York, May 23, 1962). The United Nations study employs a concept of consumption which on a global basis is similar to our own.

The United Nations estimates may be summarized as follows:[3]

Basis of Estimate	Total Consumption Non-Communist world			
	1965	1970	1975	1980[a]
Trend 1950-59				
(thousand metric tons)	6,300	9,300	13,300	19,630
Annual growth (per cent)	8.1	8.1	8.1	
Per capita trend 1950-59				
(thousand metric tons)	6,170	9,190	13,750	20,485
Annual growth (per cent)				
(6.1% per capita)	8.3	8.3	8.3	
GNP Correlation 1950-59				
Trend				
(thousand metric tons)	5,800	7,900	11,100	15,500
Annual growth (per cent)	7.2	6.8	6.9	
High				
(thousand metric tons)	6,000	8,300	11,900	17,005
Annual growth (per cent)	7.7	7.4	7.4	
Low				
(thousand metric tons)	5,600	7,300	9,700	12,980
Annual growth (per cent)	6.5	6.0	6.0	

[a] Extended by author based on per cent change 1970-75.

The trend projection is the most straightforward of the estimates. It is a semi-log straight line trend computed from the base period. Its validity hinges on the representativeness of the base period and the closeness of the fit of the trend. It has been suggested already that consumption may have been disproportionately affected in the earlier part of the period by the Korean War and its aftermath, especially in the United States. Also one may question whether exceptional performance of the economy with its attendant implications for aluminum consumption will continue in Europe. However, it is worth noting that the trend based on this decade is quite similar to that derived from longer periods—the long-term trend of aluminum consumption has been well sustained.[4]

Projections based on the trend in per capita consumption prob-

[3] *Ibid.*, pp. 30, 52, 75.
[4] The United Nations ECOSOC study shows a 40-year trend terminating in 1959 of 10.4 per cent per year for consumption of primary aluminum. If adjusted for differences in definition and scope, it would be very close to the figure for 1950-59.

ably have least plausibility. Aside from the element of possible error involved in forecasting population, there is no reason to expect consumption to vary directly with population. Vast differences prevail in per capita consumption from one country to another, and the greatest changes in population have not occurred in the countries where most of the increased consumption of aluminum has been absorbed. Such statistical relationship as may exist between world population growth and the growth of consumption does not appear to have much explanatory value. When dealing with individual countries it is useful to be able to compare the reasonableness of various projected levels of per capita consumption in the light of the experience of other countries with similar personal income and industrial structure, but as an independent forecasting device the trend in per capita consumption has little to recommend it.

Correlation techniques have been employed by many forecasters of aluminum demand. The most frequent conclusion has been that the best relationship is between consumption and gross national product (GNP). This relationship was computed for the United Nations study based on the records of individual countries during the base period and was extended, using independently available forecasts or trends of GNP growth. Depending on which GNP forecasts are accepted, the results range from as low as 6 per cent to as high as 7.7 per cent for 5-year intervals, 1960–75, with the estimates based on the trend in GNP growth at about 7 per cent.[5]

Other studies have attempted correlations with industrial production or its durables component, with new construction, output of producers' durables, consumer expenditures for durables, aluminum prices, and price ratios. The most common conclusion is that the best relationship is between consumption and GNP, although in dealing with American data, Rosenzweig finds that addition of a time trend factor to the relationship in a multiple correlation improves it, as to a minor extent do figures for prices and price relatives of aluminum and competing products.[6] All in all, however, the estimate derived by a simple GNP correlation and extended in accordance with the GNP trend seems a reasonable starting point for an estimate of world demand.

Should such a figure derived statistically be arbitrarily modified by judgment? Allusion has been made already to the possible

[5] United Nations ECOSOC, *op. cit.*, p. 75.
[6] Rosenzweig, *op. cit.*, pp. 28, 31.

biases imparted to the United Nations results by their choice of base period. These relate to the fact that the decade spanned an early period of rapid general economic growth both in Europe and North America as a consequence of postwar readjustments and the further stimulus of the Korean War in the United States. It should be added that a constricted supply of metal impaired the growth of consumption in both areas during the early part of the decade, but by the end of the period this no longer was a factor. Other causes are responsible for the slackened rate of consumption growth at the end of the 1950's, especially in North America where the economy was sluggish.

However, the growth of aluminum consumption has depended not merely upon the expansion of the economy or the restoration of prewar industry; it also has relied upon the growth of new uses or the displacement of other materials. This in turn is a consequence of changes in the product mix and of developments both within and without the aluminum industry which affect its relative advantages either with respect to properties of the materials or its relative prices.

The most obvious of such changes would be those that would lower the relative price of ingot aluminum. They might be either technological or institutional in origin, reflecting either lower costs or greater competitiveness in various markets. Less obvious would be improved techniques of working the metal via extrusion, drawing, thin rolling, etc., which might give it an advantage over competitors in fabricated form. New alloys, insulating, or coloring developments may make it competitive in new uses. Marketing innovations, such as provision of molten metal to large users or pioneering new end product applications, may also be important. Even such remote developments as alterations in building codes and the extension of prefabrication can affect the rate of substitution.

It is difficult to separate the portion of the growth of consumption which may be owed to new uses or substitution from that part representing growth of existing uses. However the United Nations study suggests that from a comparison of the differential growth rates of aluminum-consuming industries and their absorption of aluminum about three-fourths of the total consumption growth in the United States during the 1950's stemmed from new uses and substitution whereas in Europe the figure was only slightly under one-half.[7] The study readily concedes that use of the base period chosen builds a significant rate of substitution into the forecast.[8]

What are the prospects that this will continue? While the complexity of the question has been stressed, it is probable that in the forthcoming competition among materials relative prices will play a big role. Therefore, a discussion of aluminum prices seems in order at this point.

ALUMINUM PRICE TRENDS AND THEIR SIGNIFICANCE FOR DEMAND

As a comparatively new industry, aluminum's technology has been subject to constant improvement over the long sweep resulting from engineering advances and from increased scale of operation. Aluminum prices have fallen on a deflated basis from the early years of this industry, no doubt in large part as a consequence of these cost changes and perhaps also due to increased competition.[9]

This drop may be observed most easily by looking at price trends of aluminum ingot in the United States since 1900 as shown in Table 9. (While the data pertain only to the United States, trends in other major world markets, if converted to dollars at prevailing exchange rates, are roughly comparable [Table 10].) Two deflators have been used for the American prices. The first is the U. S. Bureau of Labor Statistics wholesale price index which permits comparison of movements in aluminum prices with changes in other wholesale prices. By this standard it is apparent that aluminum prices have declined quite dramatically, the deflated price being generally one-fourth to one-fifth of the 1900 level during the most recent decade. In early postwar years it dropped still lower but has tended to stabilize since then at a level of about one-half the 2 prewar decades. The second series employs a GNP deflator, and by

[7] United Nations ECOSOC, op. cit., Appendix, pp. 39-40.
[8] Ibid., Appendix, p. 38.
[9] In absolute terms, of course, the price trend is obscured by inflationary developments in different countries. However, even on an absolute basis, prices tended downward in the United States until 1948 when the general postwar American inflation and heavy demand for metal combined to reverse the trend. Nonetheless, by 1964 the price of ingot at 23.7 cents per lb was barely above the level prevailing during the depressed 1930's and still well below prices of most earlier years.

Table 9. Aluminum Ingot Price in the United States, and Trend if
Deflated by Two Series, 1900, 1910, 1920, 1930, 1940-64

Year	Price (cents per lb)	Index of Deflated Price (1900=100)	
		Using BLS Wholesale Price Index Deflator	Using GNP Deflator
1900	32.7	100.0	100.0
1910	22.2	54.1	61.8
1920	30.6	34.0	40.2
1930	23.8	47.3	39.4
1940	18.7	40.8	35.8
1941	16.5	32.4	28.8
1942	15.0	26.1	23.2
1943	15.0	25.0	21.1
1944	15.0	24.8	20.7
1945	15.0	24.3	20.2
1946	14.0	19.9	17.6
1947	14.0	16.2	15.7
1948	14.7	15.7	15.7
1949	16.0	18.0	16.9
1950	16.7	18.1	17.5
1951	18.0	17.5	17.3
1952	18.4	18.4	17.5
1953	19.7	20.0	18.4
1954	20.2	20.4	18.8
1955	21.9	22.1	20.2
1956	24.0	23.4	21.5
1957	25.4	24.1	21.9
1958	24.8	23.2	21.0
1959	24.7	23.1	20.5
1960	26.0	24.2	21.3
1961	25.5	23.9	20.7
1962	23.9	22.3	19.1
1963	22.6	21.2	17.8
1964	23.7	22.2	18.4

Sources: Prices from Metallgesellschaft Aktiengesellschaft, *Metal Statistics* (Frankfurt am Main, annual). The U.S. Bureau of Labor Statistics (BLS) deflator is from *Historical Statistics of the United States, 1957*, series E 13-24 and E 25-4, and from subsequent issues of the *Monthly Labor Review*. The GNP deflator is from *Historical Statistics of the United States, 1957*, series F 1-5, and from subsequent issues of the *Survey of Current Business*.

Table 10. Price of Aluminum Ingot in Major Markets,
Selected Years, 1930-64

	U.S. Cents per Kg				
Year	Canada	United States	Germany	United Kingdom	France
1930	—	52.5	44.4	44.2	46.4
1940	—	41.2	53.2	43.5	43.9
1950	33.0	36.8	42.4	31.4	34.3
1955	47.5	48.3	52.9	46.0	50.3
1960	51.3	57.3	51.8	51.3	47.9
1961	51.3	56.2	54.1	51.4	49.8
1962	52.6	52.6	54.0	49.9	49.8
1963	49.8	49.8	54.3	49.8	50.2
1964	52.3	52.3	54.1	52.5	51.0

Source: Metallgesellschaft Aktiengesellschaft, *Metal Statistics* (Frankfurt am Main, annual).

this standard the decline is a bit more dramatic. The price shows considerable stability in recent years at a level of about one-half that of the prewar decades. These data indicate that insofar as aluminum prices reflect cost, productivity in the aluminum industry has advanced considerably faster than for the economy as a whole during the fairly brief period of its history.

Since aluminum is an industrial raw material generally employed in the manufacture of other goods, much of the effect of price on demand will occur by way of the substitution of aluminum for other materials, and vice versa. For consideration of this question it is necessary to look at the relative prices of aluminum and its principal substitutes. In most uses the substitutes will be steel, other nonferrous metals (especially copper and zinc, and sometimes tin and lead),[10] plastics, wood, concrete, and glass.

Some comparisons of relative prices of aluminum and competing metals are made in Table 11 on the basis both of price per unit of weight and per unit of volume. Neither may be wholly appropriate for a specific purpose because functional requirements of strength, electrical conductivity, etc., may be such that aluminum does not substitute pound for pound or volume for volume with other materials. Thus, as an electrical conductor, for example, aluminum has

[10] Titanium and magnesium may deserve mention for the future.

Table 11. Ratio of Price of Other Metals to Aluminum Price in the
United States, Expressed Per Unit of Weight and Per Unit of Volume

(A= per unit of weight; B=per unit of volume)

(Aluminum Price=1.00)

	Steel		Copper		Lead		Zinc	
Year	A	B	A	B	A	B	A	B
1930	0.06	0.18	0.55	1.80	0.23	0.97	0.19	0.51
1935	0.06	0.18	0.42	1.40	0.20	0.83	0.21	0.56
1940	0.08	0.24	0.60	2.00	0.28	1.16	0.34	0.90
1945	0.10	0.29	0.79	2.60	0.43	1.82	0.55	1.45
1950	0.16	0.46	1.27	4.21	0.80	3.34	0.83	2.19
1955	0.15	0.44	1.71	5.67	0.69	2.90	0.56	1.48
1960	0.15	0.45	1.23	4.08	0.46	1.93	0.50	1.32
1961	0.16	0.46	1.17	3.88	0.43	1.79	0.45	1.20
1962	0.17	0.49	1.28	4.24	0.40	1.69	0.49	1.29
1963	0.18	0.52	1.35	4.48	0.49	2.07	0.53	1.40
1964	0.18	0.52	1.35	4.46	0.57	2.41	0.57	1.51

Source: Metallgesellschaft Aktiengesellschaft, *Metal Statistics* (Frankfurt am Main, annual).

only about 61 per cent of the conductivity of copper for conductors of equal dimension, but on a weight basis the conductivity of aluminum is about two times that of copper.[11] If one wishes to compare the amount of aluminum required to do an equivalent job as electrical conductor, then 1.65 times as much volume or one-half as much weight would be required. Other examples may prove more complex, however, for the design requirements in other uses may include structural strength, flat area coverage, mass, etc., in varying proportions, and the comparisons would be likely to require reference to the appropriate alloys or varieties of the competing materials.[12] Since aluminum is not best in uses where requirement for strength is high, it tends to compete in fields where volume is the most significant comparison. In building hardware, siding and roofing, containers, machine housings and other castings, consumer

[11] The Aluminum Association, *Standards for Wrought Aluminum Mill Product* (New York, 1962), p. 6, Tables 91-92.

[12] It has been suggested that in structural applications where strength is important the required ratio of aluminum to steel by volume is two or less. United Nations, *Competition Between Steel and Aluminum*, E/ECE/184, E/ECE/STEEL/81 (Geneva, 1964), p. 67.

durables, and similar materials, the dimensions of the end product often dictate the amount of material employed, and in such cases the more relevant comparison is that on a volume basis.

The trend of relative price changes is observable from either set of figures in Table 11, however. If the price of aluminum in each year is taken as the base and other prices are expressed as percentages of this figure, then it is quickly apparent that the price of aluminum has become more competitive relative to other metals if viewed from earlier decades. By the same standard, however, these relative price movements have tended to stabilize in postwar years in the case of steel and copper while lead and zinc have become more competitive. So far as metals are concerned, the chief competition for aluminum is provided by steel and copper, and in each case there has been no persisting alteration in their relative price in postwar years.

However, in the case of all of these competing metals, prices relative to aluminum are on the order of twice the prewar level. Such a shift might be expected to cause the substitution of aluminum for other metals and it is plausible to expect that much of this substitution occurred during the period used in computing our trend (it may still be continuing in attenuated form), for adjustment to the new pattern of relative prices cannot be completed in a short period of time. If no further major shifts in relative prices occur, then the support of this force will be withdrawn and the future trend may not match that of the postwar years.[13]

As of 1964, aluminum was not quite twice as expensive as lead and zinc on a weight basis but was well below the price of copper. Steel was about 18 per cent of the aluminum price on a per unit of weight basis. If the comparison is on a volume basis, however, then steel was about half as expensive as aluminum and all of the others were considerably more expensive. Thus, aluminum has a clear price advantage over the chief other metals, except steel, in uses where dimensions are given by design requirements. Its disadvantage vis-à-vis steel in such uses has narrowed to the point where in a growing number of instances other desirable features may tilt the balance in favor of aluminum despite steel's continuing price edge.

[13] Few forecasters have paid explicit attention to this possibility. An exception, however, is Merton J. Peck, *Competition in the Aluminum Industry* (Cambridge: Harvard University Press, 1961), pp. 160-62.

How sensitive is aluminum demand to price change or to changes in prices relative to other materials? These questions have been investigated in other studies and, though the techniques employed were quite varied, the results may be reported briefly here. There appears general agreement from statistical studies that demand for aluminum is not very responsive to short-term changes in prices. One study suggests a price elasticity of –0.3 based on a deflated price series for the years 1919–54.[14] Another produced a similar though qualified result of –0.43, again using a deflated series.[15]

It might be expected that demand would be more responsive to price over a long period since this would allow time for adjustment to the changed price pattern through increased awareness of the changed possibilities, necessary retooling, and public acceptance of the new uses. Thus, employing a somewhat different approach, Peck[16] has concluded that the long-run price elasticity of demand for aluminum is –1.15. Peck reinforces this conclusion by reworking survey data compiled by Engle, Gregory, and Mossé which suggest a long-term price elasticity of –1.8 for aluminum.[17]

Any such long-term price elasticity, of course, is apt to reflect considerable substitution. The question of sensitivity to change in price of competing materials was considered explicitly in at least two instances. Rosenzweig found aluminum demand not greatly affected by changes in the aluminum–copper price ratio and only moderately influenced by changes in the aluminum–steel ratio.[18] Peck, however, suggests that there is a strong long-run relationship between aluminum demand and the aluminum–steel price ratio in particular, the cross elasticity being on the order of 2.[19]

Plausibility supports the view that over the long run aluminum demand will be sensitive to the prices of competing materials. In few cases are the properties of aluminum so unique that it does not face competing materials and, if it is to win new and hold existing

[14] Rosenzweig, *op. cit.*, p. 29. This simply means the ratio of change in consumption of aluminum that accompanies a given change in its price. Thus, in this instance, for a 1 per cent increase in price there would be a 0.3 per cent decline in consumption.

[15] F. M. Fisher, *A Priori Information and Time Series Analysis: Essays in Economic Theory and Measurement* (Amsterdam: North-Holland Publishing Co., 1962), p. 109.

[16] *Op. cit.*, p. 31.

[17] *Ibid.*, p. 33.

[18] *Ibid.*, p. 27.

[19] *Ibid.*, p. 32.

markets against this challenge, its relative price will be an important factor. Failure to improve on existing price ratios may be expected to slow the growth of aluminum consumption to a rate somewhat nearer the growth of the general economy.

The pattern of relative prices of aluminum and of competing metals already has been noted. Long-term improvement in the price competitiveness of aluminum has not been sustained for the past decade or more.[20] This becomes of particular importance in the competition with steel with respect to which aluminum remains the more expensive material on both a weight and volume basis. It is at the expense of steel perhaps more than the other metals that aluminum must hope to gain new markets. What are future prospects for relative price movement?

Over the longer run, price is likely to reflect changing cost relationships. Subsequent discussion will examine the prospects for aluminum technology and some of their implications for costs. It is enough to say at this point that there is room for considerable improvement within the context of existing technology, and alternative technological possibilities stand a good chance of adoption during the period of interest. If inputs are valued at current prices, this suggests the possibility of cost reduction for ingot produced at new plants of 15–25 per cent during the period of concern. Meanwhile, however, steelmaking technology will not be stagnant. Improved heat rates, the use of oxygen injection and oxygen converters, direct reduction possibilities, and improved feed materials to blast furnaces are among the possibilities open to the steel industry.[21] It is impossible to forecast accurately the speed of these changes, but on balance it seems quite likely that steel will keep pace with aluminum in cost reduction over the period of concern and it could do even better.

Beyond the ingot stage, developments in milling can be expected in both industries. Although one of the intrinsic advantages of aluminum is its easy workability, heretofore aluminum has been at a

---

[20] Price developments since 1964 have been markedly favorable to aluminum, especially vis-à-vis copper. Short-term factors, however, appear to predominate in this shift.

[21] H. H. Landsberg, L. L. Fischman, and J. L. Fisher, *Resources in America's Future* (Baltimore: Johns Hopkins Press for Resources for the Future, Inc., 1963). The authors discuss prospects of steel technology (pp. 212-14) and suggest at one point that the industry is probably on the threshold of major changes (p. 213).

disadvantage at this stage because the tradition of customized work and the small size of orders have added greatly to the cost of basic aluminum mill shapes. Greater standardization and the adoption of newer techniques such as continuous casting may help to cure this problem.

All in all, probably the safest assumption to make concerning the future aluminum–steel price ratio is that it will remain about constant. There appears little basis for assuming that the price of aluminum will decline much relative to steel. Thus, in its quest for new markets at the expense of steel, aluminum will be relying more upon exploitation of the intrinsic qualities of the material and adapting to a changing product mix than on further cuts in relative price.

As was suggested earlier, aluminum competes with many other materials, particularly in the packaging and building materials fields. It is even more difficult to make direct comparisons on a weight or volume basis with such diverse materials as plastics, glass, wood, or concrete. Yet it is possible to suggest that such recent inroads as aluminum has made in markets held by these materials have not been the result of favorable relative price shifts but rather have stemmed from continued exploitation of the basic wartime shift in relative prices and from further recognition and development of the intrinsic advantages of the material.

This may be seen from a comparison of price-index movements in the United States. During the period 1939–47 while the price of aluminum decreased 30 per cent and the price index of nonmetal structural minerals, including cement, glass, etc., went up 35 per cent, the lumber price index increased 197 per cent. But the trend is not nearly as favorable if postwar years are considered. Thus, during the period 1947–64 the price index for lumber rose by 30 per cent and for nonmetal structural minerals by 47 per cent, while aluminum prices increased 69 per cent.

Of even greater interest is the trend in plastics prices. Plastics were not very important prewar. In the postwar years 1947–63 the price index for plastics declined slightly in contrast to the 62 per cent rise in aluminum prices.[22]

Plastics provide major competition for aluminum in a host of uses such as building materials, packaging materials, and as housings

[22] Price indices from U.S. Bureau of Labor Statistics, *Monthly Labor Review*, various issues.

or casings. They are relatively new materials whose potential uses have by no means been fully explored. They are abundant, the product of a dynamic technology, and are more than likely to be the aggressor in future competitive struggles for market position, backed by a further decline in relative price. Therefore, those comments made earlier concerning the significance of the arrested shift in relative price of aluminum and other metals apply with particular force to its competition with plastics and to a lesser degree to the other materials with which aluminum competes. Future substitution in response to price change is unlikely to favor aluminum as it often has in the past, and its most vigorous competitor (plastics) enjoys a position vis-à-vis aluminum that is analogous to that which aluminum so long enjoyed with respect to other metals—namely, a declining relative price and a growing industrial awareness of its potential uses.

Few studies or demand projections have given much attention to the possible significance of this altered climate of relative prices for growth of aluminum demand. However, it is useful to refer again to the United Nations estimate that for the United States in the 1950's, 72 per cent of the growth in aluminum consumption stemmed from substitution or new uses and only 28 per cent from the growth of consuming industries, while for Europe the figure was about 47 per cent[23] from new uses and substitution. Price is obviously an important consideration in the further exploitation of this major component of growth in demand. A sustained period of stable relative prices should diminish the ease with which aluminum substitutes for other materials.

In the light of the foregoing discussion concerning relative prices which can be expected to be less favorable to aluminum in the future than their effect is believed to have been during the base period, it would seem logical to lower the expected future relationship between aluminum consumption and growth in GNP. With one-half to three-fourths of the base period growth attributable to new uses and substitution and therefore vulnerable to less favorable relative price trends, the potential decline in growth rate

[23] United Nations ECOSOC, *op. cit.*, Appendix, pp. 39-40.

is considerable. But it is difficult to evaluate the extent to which other developments may prove favorable to aluminum and offset the effect of price trends on substitution.

This becomes a matter of judgment. It depends in part on the growth of industries where most aluminum presently is used. If such industries grow faster than the general economy, the rate of growth of consumption will still exceed that for GNP even in the absence of any net substitution. In industrial countries, this involves speculation in particular about the automobile industry, construction (especially the United States), and, to a lesser extent, machinery packaging, and electrical equipment. In nonindustrial countries attention must be focussed more on the electrical industry and on unexploited possibilities in construction and transportation. To be realistic, however, availability of foreign exchange and priorities in its use also must be borne in mind in these cases.

It is easy to foresee fairly rapid expansion in Europe. The European auto industry, for example, has been a rapidly growing one with a strong domestic market and good export record. Aluminum is well established as a material in European cars. The picture is less promising for autos in the United States where the auto market is more nearly saturated and aluminum has not won acceptance in engine blocks. In the construction field the use of aluminum is much further advanced in the United States, and a prospective housing boom during the period of forecast suggests a strong market for this use in America. While consumption for this purpose should expand in Europe, the application of aluminum in this use is less well developed. However, as Europe shifts from emphasis on mere replacement of its wartime housing deficit to better housing quality, aluminum may find a greater market. Aluminum is unlikely to become extensively used as a building material in nonindustrial countries in this period even if building should become relatively more important, but in the specialized form of roofing sheet it may achieve some success. There have been great hopes for expanded use in packaging, and the prospects appear similar in Europe and North America, i.e., the use of packaging will grow somewhat faster than the general economy. This use will be unimportant in nonindustrial countries unless they have a specialized consumer export requiring packaging. Electrical industries are likely to grow faster than GNP in industrial countries. In nonindustrial countries, electrical applications may expect fast growth and should prove a major market. Another significant possibility in such countries is the growth of transportation industries other than autos, such as motorcycles, bicycles, pedicabs, and self-propelled rail cars, which may consume greater amounts of aluminum. On balance the changing product mix among industries now using aluminum should remain favorable to aluminum—perhaps more so in Europe than in North America—and distinctly so in nonindustrial countries where consumption nonetheless will remain very small.

It is more difficult to discern the possibilities in new uses. In the industry these hopes tend to focus around new applications in transportation, construction, and packaging, along with additional penetration of electrical industries and ordnance.

Aluminum has been used on special passenger trains and ships for some time but has failed to develop a major market. However, for the lighter weight, quick acceleration cars of urban transportation systems, for buses, ground-effect machines, and hydrofoils, it has many advantages, and there are good prospects of advance in these areas. Likewise, in nonindustrial countries aluminum may prove advantageous where transportation depends on muscle power or where roads and rail beds favor lighter equipment. A promising major market is in rail freight cars where, as has been shown in the case of hopper cars, weight savings can significantly increase capacity. Aluminum already has major penetration into the truck-body field and is gaining favor in truck tankers where weight and corrosion resistance are important. Some hold high hopes for aluminum pipe, but it lacks strength to withstand pressures in larger dimensions at competitive gauges. Therefore, its application is apt to be limited to special-purpose uses. As oil-drill pipe, however, aluminum appears to have considerable promise and it also will continue to expand in sprinkler irrigation systems. While it is not a new use, one must speculate that mobile homes—now mostly aluminum—will continue to find favor in the United States and may gain a foothold in Europe and elsewhere. On the debit side, aluminum is exposed to loss of some of its market in aircraft as speeds push beyond Mach 2.5.

New applications in construction promise to be an expanding market but the advance may be strongly challenged. Existing penetration in doors, windows, screens, hardware, conduit, gutters and downspouts, siding, etc., must resist the new advances from polyvinyl-chlorides and a more determined threat from steel. Best prospects are for more aluminum siding and greater use in wiring, while sandwich panels of aluminum and other materials will have applications in both commercial and residential construction. The slowness with which new building methods are accepted has impeded aluminum's penetration into this market and by the same token may help to protect its position against assaults from plastics and steel.

Packaging applications have been the focus of much research and marketing effort, especially by American firms. Lightweight, attractiveness, the ease with which it can be printed on in color, resistance to corrosion, and easy formability have favored aluminum. However, many of the same features are found in plastics with the often desirable added feature of transparency. Aluminum has penetrated a few can markets based on its characteristics, notably the American fruit juice market, but it remains an expensive material. Reduction in fabrication cost may partially solve this problem and real or imagined advantages (pop-top beer cans, for example) will bring added applications. However, the potential does not seem to warrant the hopes of some in the industry. There remains the haunting spectacle of aluminum's fight for the American oil can market where a successful campaign against steel soon turned to ashes as a foil-fiber product emerged from the fray with the prize while a plastic-fiber product threatened still further inroads on the aluminum penetration.[24] Thus, once packagers decide to take a fresh look, they may come to rest with a product quite different from that which the challenger proposes. Yet another case in point is the aforementioned fruit juice concentrate field where the foil-fiber unit again may displace aluminum.[25] Despite all of the commotion about cans, the aluminum industry is likely to make only a marginal penetration in specialized uses, plus a greater share of the market for can ends where ease of opening is a significant factor.[26]

[24] *Journal of Commerce*, May 3, 1963 (New York), pp. 1, 23.
[25] *Ibid.*, April 24, 1963, pp. 2, 5.
[26] An intriguing suggestion has been made that beer keeps significantly longer in aluminum cans without acquiring a metallic taste, and therefore it would pay brewers to flatten out the seasonal pattern of their production by producing for inventory in aluminum cans. This would be very significant if it should prove true, because of the size of the market. This, however, may presume an unwarranted sensitivity of the beer drinker's palate.

In the electrical industries aluminum has only limited fields to conquer; new transmission and above-ground distribution lines already are almost exclusively aluminum. Household wiring and underground distribution plus certain engineering uses offer the remaining possibilities. The greater bulk of the material required when compared to copper has been the drawback here for it compels use of more expensive insulating materials. Conceivably cheaper insulating materials may accelerate the penetration of aluminum into the underground distribution field, but the prospects for household wiring seem less immediate. In motor windings, the development of strip windings has helped to make aluminum more competitive but, because of its compactness, copper will retain much of this market. Of course, if copper prices hold at the level established in late 1965 and early 1966, the longer-run opportunities for aluminum to penetrate additional electrical uses will be great.

The increased focus on military mobility, particularly in the United States military establishment, is favoring new uses of aluminum since it requires that equipment be light enough to be transportable by air. New tough aluminum alloys have been developed for use in armor plating, and it may be expected that a diligent and systematic evaluation will be made of military equipment to learn where aluminum can substitute for steel. With the metal cost normally only a small percentage of the total cost of military equipment, this is the sort of use where a premium can readily be paid for savings in weight. By the same token, one of the most important military uses of aluminum—military airframe manufacture—faces a very uncertain future as speeds advance beyond present Mach numbers. Missile airframes and propellants will continue to absorb metal, but this does not promise to be a truly large-scale market.

The foregoing is only qualitative in nature. It suggests that there are many applications where aluminum can find new or expanded uses. A thoroughgoing end-use study by country would be needed to quantify these possibilities, and such an endeavor is beyond our scope.

PROSPECTIVE DEMAND FOR ALUMINUM IN NON-COMMUNIST WORLD

Sketchy as the foregoing has been, it does appear to support the view that aluminum consumption should continue to grow faster than real GNP during the period ahead as a consequence of favorable developments in the product mix in existing uses and from the development of new uses. Yet because the support of favorable trends in relative prices reflected in the consumption of the base period is not likely to continue, the rate at which aluminum substituted for other materials during the base period may not be sustained. This suggests that some reduction in the 1950–59 trend is possible. As a maximum, it is unlikely that the annual rate of growth of consumption would lag as much as 2 percentage points from the figure derived from the base-period relationships. Thus, at most the rate might fall as low as 5 per cent per year instead of the 7 per cent based upon GNP trend; in that case, consumption would grow about one-third faster than GNP instead of 80–100 per cent faster as in the past. For the non-Communist world, growth at a straight 5 per cent per year projected from 1964 would yield:

	(million metric tons)
1970	8.0
1975	10.1
1980	13.0

These figures may be compared with the alternatives from page 55 to determine the range of what might appear reasonable estimates. Thus, we have a possible 1980 range of from 13.0 million metric tons based on the suggested 5 per cent growth factor to as high as 20.5 million metric tons if per capita growth rates shown in the United Nations study are extended. As a working range of possibilities, we will consider a 5 per cent figure as the lowest, 8 per cent as the highest, and 7 per cent as the most likely figure. Hence the following range of possibilities extrapolated from 1964:

	(million metric tons)		
	5%	7%	8%
1970	8.0	8.9	9.4
1975	10.1	12.5	13.8
1980	13.0	17.5	20.3

The foregoing allows for considerable local variation, with use spurting ahead in some countries and growing slowly in others. How should these figures be allocated regionally? Although the United Nations figure used was originally based on correlations with GNP changes by country, each extended by the trend of GNP

71

growth, the United Nations was chary of publishing this detail, perhaps because the results for some countries seem quite improbable.[27] The implications of several alternative procedures for regional allocation of the global projection are given in the appendix to this chapter.

None of the statistical procedures for estimating future consumption regionally inspires confidence. Forecasts from diverse sources for various countries and continents were reviewed, but they displayed wide discrepancies, even when differences in concept were allowed for. In the final analysis no procedure is likely to yield unquestioned results and informed judgment may serve as well here as any other method. Accordingly, our most probable figure for 1980 has been distributed in an arbitrary manner in which past growth rates, per capita income, and stage of development were considerations. The distribution supposes a more rapid growth of consumption in nonindustrial countries. Consumption in Europe is assumed to grow somewhat faster than North America, but the rate and the advantage over North America will diminish from the base period. The same is true of Japan, but the rate of growth will remain notably higher than in Europe and North America. North America, while the slowest growing market, nonetheless will more nearly approach the diminished European rate. Asia (other than Japan), Africa, and South America may absorb much greater amounts of metal even though in total they may use less than one-tenth of the 1980 non-Communist total. Oceania also should increase its share sharply, reflecting consumption more nearly in line with income, but it will remain a small market. The figures reflecting these assumptions are shown on the opposite page.

The figures for Japan and Asia are perhaps the most tenuous since, although the base-period growth in Japan has been so extraordinary, Japan still has not reached a relatively large figure for an industrial country. The aspirations of other Asian countries are high but consumption is limited by shortage of foreign exchange,

---

[27] It may be asked why the distribution is not simply allowed to reflect the results of the computations by country which were used to arrive at the total. The reason is that for some countries the results are extravagantly high while for others the base-period data were too skimpy to provide any real basis for forecasting. In particular, Germany and Japan cannot be expected to continue their trend of the 1950's, and for most of Latin America, Africa, and Asia the data are very poor. The total, however, is plausible in view of the results of other forecasting techniques.

Total Projected Consumption of Aluminum
(Excluding the Communist Bloc), 1980

	Consumption (thousand metric tons)	Implied per capita (kg)
Europe	6,100	16.1
Latin America	550	1.6
United States and Canada	8,400	33.0
Asia	1,950	1.4
(Japan)	(1,400)	(11.4)
Africa	200	0.6
Oceania	300	13.3
Total	17,500	6.3

especially in India. Success in their industrial programs might increase their consumption because of greater domestic production or enhanced availability of exchange. Under such circumstances their consumption would be still higher. However, under the best of assumptions the per capita level of consumption in Asia outside of Japan will remain low—probably lower than Africa because South Africa will consume metal at the rate of an industrial country, and perhaps no higher than the rest of Africa where the demands of electrification and the utility of aluminum for tropical construction will combine with a happier exchange outlook to result in significant growth of consumption. The comparative optimism for Latin America assumes further industrialization and economic integration in Latin America that will favor domestic production and consumption more nearly appropriate to their income levels. The figures for United States–Canada allow for an average growth rate from 1964 of about 6.2 per cent per year while those for Europe permit an average rate of 7.0 per cent. These figures also imply an average Japanese growth rate of 8.7 per cent.

PROSPECTIVE AMOUNT OF SECONDARY METAL

If the foregoing is accepted as the approximate dimension of future total consumption, it is still necessary to estimate the portion of this which will be primary metal. It is assumed that secondary metal will be employed to the extent available, with residual demands net from primary production.

While it was necessary to consider total consumption on a regional basis, there was no need to carry the distinction between primary and secondary consumption to that level. However, on a global basis, the approximate ratio between the primary and secondary components must be estimated, recognizing that this allows for wide variation among individual countries whose choice between domestic production of either kind and imports of either kind will depend upon local conditions.

The nature of the secondary market is such that comparatively little secondary metal moves in international trade in refined form.[28] There is a fairly active international trade in scrap, but for the most part a country that seeks to produce its own metal will rely upon domestically produced primary and on domestic scrap. No major aluminum-consuming country relies upon converting imported scrap as the basis of its metal supply.

Secondary metal is produced from scrap originating in two broad sources: new or current industrial scrap generated from the fabrication of metal and old scrap recovered from finished products previously in use. The supply from each source has its own set of determinants, but in general the supply of current scrap is a function of the volume of current production in metal-using industries and of the techniques employed, while the supply of old scrap depends on the use cycle of metal-containing products, the past rate of absorption into such uses, and the cost of collection and recovery.

Theoretically, it would be possible to derive industrial scrap generation factors for various metal-using industries and apply these to an estimated future distribution of end uses to arrive at a total figure for industrial scrap that might then be further adjusted by a factor for the portion of it recovered. Such a procedure would assume that the techniques of metal use and share of recovery would not alter greatly. However, the data provide only the most tenuous basis for this procedure in the United States and no basis for most other countries.[29] Moreover, since we have made no projection of end use on a global basis, this procedure is not adaptable. Instead it will simply be assumed that a fixed ratio of recovered industrial scrap is generated by the industrial process. Based on

[28] Most is oriented to local supplies of scrap and to local customers, generally founders, die casters, and steel mills.
[29] Landsberg *et al., op. cit.*, have worked out such figures for the United States. Their concept of consumption is net of the industrial scrap generated in each use (p. 900).

American experience of the past decade, this ratio appears to be in the range of 12–15 per cent of total consumption.[30] The ratio, of course, is subject to change in response not only to changes in the mix of uses and techniques of use but also the demand for scrap and the ease of collection. If any trend is discernible in the ratio, it appears to be diminishing—perhaps in response to more efficient use of metal at the fabrication stage.

Fairly elaborate techniques have been devised to estimate the future supply of metal from obsolete scrap. They depend upon estimating the past and expected flow into various uses, the length of time the metal, on the average, will remain embodied in these products, and the percentage of recovery at the end of the period (Table 12). Again, it is apparent that to apply such a formula it is

Table 12. Illustrative Table Showing Percentage of Recovered Obsolete Scrap to Total Consumption Under Varying Assumptions of Length of Recovery Cycle, Recovery Ratio, and Rate of Growth of Consumption

Length of Recovery Cycle (Years)	Per Cent of Recovery	Rate of Growth of Consumption			
		5%	6%	7%	8%
		Recovered Obsolete Scrap as Per Cent of Total Consumption			
15	40	19	17	15	13
	60	29	25	22	19
	80	38	33	29	25
20	40	15	12	10	9
	60	23	19	16	13
	80	30	25	21	17
25	40	12	9	7	6
	60	18	14	11	9
	80	24	19	15	12

necessary to have good end-use data, both past and future; therefore it is not usable here. However, if the implications of these relationships are worked out it is clear that recovery from old scrap is considerably more important relative to total consumption if the growth rate of consumption is low and also if the recovery cycle is

[30] See Metallgesellschaft Aktiengesellschaft, *Metal Statistics* (Frankfurt am Main, 1964), p. 76. United Nations ECOSOC, *op. cit.*, p. 88, suggests a 12 per cent figure. It is difficult to be precise because the United States is an exporter of scrap which is not segregated between new and old material.

short and the ratio of recovery high.[31] Aluminum has maintained a high growth rate and even under the most adverse conditions is expected to continue to have a fairly high rate of over 5 per cent per year. The recovery cycle varies from an extremely short period in the case of containers to as long as 50 years in electrical transmission and perhaps equally long in construction. In the important transportation category it is probably closer to 10 years, perhaps the same in consumer durables, and double this in producer durables. There is a tendency to consider 20 years as a fairly good weighted average life in use.[32] The proportion which is recoverable also varies widely among uses. Metal available in large chunks or where channels of recovery are well developed would be expected to show the best recovery. Thus, aircraft, producer durables, auto engines, electrical transmission cables, and bus bars should have a high recovery ratio. Metal used for construction and consumer durables would be in a more dubious category, while that used in containers and certain outright destructive uses as in paints, rocket fuels, and deoxidizers will have no recovery. Unfortunately, we have no data by which we can determine actual recovery ratios. It may well be that aluminum is tending to move into uses where recovery ratios will be low. Containers are an obvious example. Also, the increasing uses in window frames, nails, electrical conduit, and building bespeak a long and uncertain cycle of recovery, while even in automobiles much of the aluminum is going into trim and castings that can be segregated only at a rising cost.

Those who have applied formulae as an approach to estimating recovery from old scrap have tended to arrive at fairly sizable figures. Thus, Landsberg et al.[33] offer a medium figure of 680,000 metric tons for the United States by 1980, while United Nations ECOSOC[34] suggests (but does not insist upon) a United States figure of as high as 1 million tons by 1975. Under plausible assumptions of growth rates, recovery ratios and life cycles in use in the application of a formula would suggest possible recovery from

[31] United Nations ECOSOC, op. cit., p. 92, makes these points and works out a table showing various possibilities.
[32] Ibid., p. 96, suggests such a figure.
[33] Op. cit., p. 902. They assume a rising percentage of recovery from a computed "potential supply." However, the record which they cite, 1950-60, showed a very sharply diminishing percentage of recovery from that computed figure—from 50 per cent to only 32.1 per cent.
[34] Op. cit., p. 96.

old scrap of 3–4 million tons by 1980 on a global basis or perhaps 15–25 per cent of the world total.

When one looks at the historical data on secondary production, however, there is no apparent tendency for the ratio to increase. On a global basis total secondary production, which was 26.1 per cent of the consumption in 1948, drifted downward and stabilized at about 20 per cent during the period 1952–64. It averaged closer to 25 per cent in Europe but showed no trend. In the Americas the figure fell from 26.7 per cent in 1948 and has been on the order of 15 per cent for most of the past decade.[35]

These historical figures do not distinguish between new and old scrap. In the United States, using a somewhat different series, old scrap has supplied 3–5 per cent of total consumption over the 1952–64 period and new scrap 12–16 per cent, with no very clear trend but a slight downward drift.[36] This relationship was maintained despite the depressed consumption of part of the period. It could be explained by a diminishing ratio of recovery, as suggested by the Landsberg *et al.*[37] data, or by improved (metal-saving) techniques of use which would diminish the potential supply of new scrap.

The American figures suggest that about 75–80 per cent of all secondary metal currently comes from new scrap, but on a global basis the ratio may be lower—perhaps two-thirds. This would imply that about 13–14 per cent of total world consumption now comes from new scrap. Allowing for added economies in metal use, a ratio of 12 per cent or less for new scrap by 1980 appears reasonable.[38] Thus, 2.1 million tons might be supplied from this source.

Old scrap presents greater difficulty. Should we look at the skimpy record or rest our faith in a theoretical construct? The results will be very different. Because there is reason to suspect that the American figures at 3–5 per cent of total supply understate the supply of old scrap (exports diminish it), they must be taken with caution. But in the light of this record, it is very difficult to accept the full implications of a computed future recovery of old scrap, on the order of 15–25 per cent of total consumption.[39] A slower

[35] The United States generates more scrap than this, however, and exports considerable amounts to Germany, Italy, and Japan.

[36] See Metallgesellschaft Aktiengesellschaft, *op. cit.* (1965), pp. 4, 5, 76.

[37] *Op. cit.*

[38] United Nations ECOSOC, *op. cit.*, p. 88, suggests this figure.

[39] See Table 12 for an illustrative computation of possibilities.

rate of growth in future consumption implies a higher ratio of re-covered old scrap to total consumption while increased costs of collection and possible deterioration in the ratio of recovery from the potential supply may act in the opposite direction. It seems best to allow the ratio of recovered old scrap to total consumption to grow from its apparent current figure of perhaps 5–6 per cent, but not to its full theoretical potential. A ratio of about 14 per cent of total consumption for 1980 is suggested, implying 2.5 million tons by that date. It should be recognized that this figure is a matter of judgment and the true result could depart considerably from it.

Combined with the previous figure for new scrap, these estimates imply that approximately 4.6 million tons of total consumption would be met from secondary metal or a little more than one-fourth of the total, up from less than one-fifth at present. By subtraction, the resulting consumption of new primary would be 12.9 million tons by 1980 compared with 4.8 million tons in 1964.

APPENDIX: RESULTS OF ALTERNATIVE METHODS OF ARRIVING AT
REGIONAL CONSUMPTION PROJECTIONS

Various procedures occur for making regional allocation of the global demand projection. Why not simply extend the growth trend of consumption by country and sum according to the regional groupings desired? (Note that this would result in a different total while at the same time allocating it.) The chief drawback of this method is that during the base period countries have been at different stages in the development of aluminum consumption, and extension of the trend to 1980 can give grossly distorted results. For example, if the average growth rate is computed for the period 1954–64, the growth rate of consumption in the United States and Canada was 7.3 per cent, in Europe 8.6 per cent, and in Japan 17.5 per cent. Extending these rates from a 1964 base would give the improbable figure of over 4.8 million metric tons for Japan by 1980.

Suppose we assume that the distribution of world consumption will not change—only its total will alter within the limits of the United Nations projection. On that basis, North America retains its lead and the United States–Canada figure would be 9,450,000 metric tons in 1980, Europe 6,055,000 metric tons, Japan 1,085,000 metric tons, other Asia and Africa 350,000 metric tons, rest of

Americas 315,000 metric tons, and Oceania 245,000 metric tons. Ratios of per capita consumption would alter slightly in response to differing rates of population growth in company with the common growth rate for consumption. However, per capita figures in United States–Canada would remain more than twice as high as European and Asian. African and Latin American per capita figures (except for Japan) would continue very small. This distribution is not very interesting, however, except as setting one kind of limit. There is no basis for its assumption that the same growth rate will prevail in all regions, and the growth rates in fact have been quite different, as was suggested earlier. This is reflected in the changing distribution of consumption. Japan, which took a negligible share of the non-Communist total in 1948, had 6.2 per cent by 1964, while the United States–Canada share was declining from 62.4 per cent to 54.0 per cent. The assumption of a fixed distribution or single growth rate for all countries may be thought of setting a likely upper limit on the United States–Canadian figure and a lower limit on Asian, African, Oceanian, and Latin American consumption.

Another possibility would be to focus on the distribution of changes in a recent period and to project the results if this relationship prevails into the future, given our limits for the global figure already stated. On this basis those areas with a large share of the change relative to their share of the total would continue to increase in relative importance. However, as their share of the total grows while their share of the change remains fixed, there is a tendency for the two figures to converge, i.e., for the share of the total to approximate the share of the change. Over time this means a convergence of growth rates towards a common figure with a new distribution of the total approaching the distribution of changes in the base period. Again, it is instructive to work out the implications of such a procedure. Using 1954–64 as a base period, it is clear that because the United States–Canada grew slowly in relation to their share of total consumption, and all other areas faster, the distribution shifts away from North America. This is especially true for Japan and other Asia and for Oceania. On this basis, and using the 17.5 million ton figure for 1980 established previously, European consumption (in thousand metric tons) would be 6,100, United States–Canada 8,800, Japan 1,465, other Asia and Africa 460, other Americas 335, and Oceania 340. Meanwhile the growth rates would converge. On a per capita basis, wide discrepancies would remain. Again this procedure is easily assailed as depending solely on the distribution of changes during the base period and thereby extending too far into the future the extravagant growth rates of Japan, while the comparative sluggishness of the American economy during much of the period likewise becomes embedded in the system.

Still another approach might be to compute an intercountry relationship between per capita consumption and per capita income, and project this relationship on the basis of assumptions concerning real income growth per head and population. Again, the procedure arouses suspicion. As of the date the relationship is computed, consumption may be in a state of transition in some countries and may not be at its "normal" level because of temporary factors. Also, given the degree of industrial specialization among countries, the relationship on a country basis cannot be a very close one, although if larger areas are used this may average out to some degree. Finally, per capita income converted at conventional exchange rates may be a very imprecise measure of real domestic buying power. Nonetheless, such a relationship was attempted for the years 1953 and 1961. Almost no correlation was found for 1953. There are many possible explanations, but certainly the insufficient time elapsed since World War II would mean that many distortions could be expected. By 1961, with exchanges freer and supply more adequate, countries were in a better position to consume at a normal level, and there was a relationship such that, for the countries considered, about 79 per cent of the difference in per capita consumption was associated with differences in per capita income. More significantly, when countries were grouped by continent (Japan treated separately), thereby smoothing out some of the distortion resulting from intraregional specialization and trade, the relationship was closer still. European consumption is somewhat above what might be expected according to the regression line, but this is not so surprising in view of its high degree of industrial specialization.

Direct application of this relationship would yield a low figure for projection purposes since it implies that a constant rather than a growing share of GNP is spent on aluminum. However, adjustment could be made to our global estimates on a proportional basis. The results then would depend upon the assumed rates of growth of per capita income. The range of possibilities here is such as to discourage further pursuit of this technique. It is interesting, however, to note that, in contrast to our previous illustrative system of allocation, this system preserves the predominance of North America in total consumption because of its high absolute level of income and its relatively rapid population growth, while Europe, even at rapid assumed growth in per capita income, is retarded by its lower absolute level and slower population growth. The dramatic gainers would be Latin America in particular (rapid population growth coupled with higher incomes than in other nonindustrial countries) and to a lesser extent, Africa. Japan would not perform as well as by the previous method except at sustained per capita income growth rates of about 10 per cent or more. While this procedure is dubious on many counts, it is useful in pinpointing what appear to be underdeveloped markets in relation to income and it puts us on notice that despite past trends such markets may be candidates for development.

CHAPTER FOUR

BASIC TECHNOLOGY OF THE ALUMINUM INDUSTRY

The basic technology of the aluminum industry has remained fundamentally unchanged from the time the Hall-Héroult process for reduction was invented in 1886 and the Bayer process for production of alumina was discovered in 1888. The efficiency of all phases has improved and a further step sometimes is added to the Bayer process to treat lower quality ores economically, but these do not represent a sharp change in the industry's technology. While there are prospects now that such changes may occur, their likelihood remains uncertain. Therefore, it is appropriate to describe briefly the existing technology and its significance for the industry's location.

Unlike some metals, aluminum is never found in nature in metallic form. It is a metal without any history prior to the age of science, and its discovery and large-scale production derive from the scientific and technical revolution of the nineteenth century.[1] After expensive efforts to produce the metal by other means, researchers working independently in the United States and France simultaneously discovered the basic reduction process now in use. Meanwhile, an inexpensive process for preparing the raw material, alumina, had been developed and aluminum began its remarkable penetration into industrial uses. This story is told in detail elsewhere and need not be repeated here.[2]

Aluminum is a common material in the earth's crust. Alumina (the oxide of aluminum) constitutes about 15 per cent of igneous rocks. Although aluminum is the most common of all metals and is

[1] See D. H. Wallace, *Market Control in the Aluminum Industry* (Cambridge: Harvard University Press, 1937), Appendix A.
[2] *Ibid.*; see also Junius Edwards, Francis Frary, and Zay Jeffries, *The Aluminum Industry and Its Production* (New York: McGraw-Hill, 1937), Vol. I.

second only to oxygen and silicon among materials in the earth's crust, it is invariably combined with other elements.[3]

While it is possible to separate aluminum from many types of ore, the easiest to work is bauxite in which aluminum is chemically combined with oxygen and water. Therefore, the first step in producing aluminum is the discovery and mining of bauxite (or, if necessary, other aluminous materials). The bauxite is refined into alumina by means of the Bayer process. With present technology this step is necessary to prevent impurities from being carried through to the metal. Alumina, in commercial form a nearly pure chemical of aluminum oxide, is the material which is charged into the primary smelters. The aluminum metal is made by separating the metal from its oxide electrolytically in large metal "pots" or cells through which DC electric current is passed. The metal is recovered in molten form and thereafter is formed and worked in much the same fashion as other metals, allowing, of course, for its special properties. Thus, it is customary to distinguish three stages in the production of primary metal—bauxite, alumina, and aluminum—with fabrication, a final stage common to secondary production as well. Each will be considered in turn.

BAUXITE: CHARACTERISTICS, MINING, AND PROCESSING

Bauxite is the only commercial aluminous material used for aluminum production today under normal economic conditions.[4] A very small amount of other ore has been used in Western countries under special circumstances, but the only significant nonbauxitic material employed at present is nepheline which is used in the Soviet Union.[5] Thus, while techniques are known for the recovery of alumina from nonbauxitic ores and may subsequently prove economical, at present they are not economical under most circumstances.

[3] United Kingdom Overseas Geological Surveys, *Bauxite, Alumina and Aluminium* (London: H.M.S.O., 1962), p. 2.
[4] United Nations Secretariat, *Pre-investment Data on the Aluminum Industry*, ST/ECLA/Conf.11/L.24, January 28, 1963, by Jan H. Reimers, p. 12.
[5] This is a special case, however; the alumina is recovered as part of a joint process. The U.S.S.R. is poorly supplied with bauxite and it pursues a policy of comparative self-sufficiency which makes it more reluctant to import raw materials than other similarly situated countries. *Ibid.*, p. 19.

Bauxite consists mainly of hydrates of aluminum. It is believed to have been formed by the weathering of rocks in areas of heavy rainfall. Such weathering breaks down the aluminum silicates found in most rocks and under certain conditions leaches out soluble constituents, concentrating the hydrated aluminum oxide which remains, along with varying amounts of impurities.[6] Since suitable conditions appear to occur more frequently in tropical climates, the bulk of the world's known bauxite reserves is found in tropical areas.

The two principal types of bauxite are monohydrates and trihydrates. The monohydrates ($Al_2O_3 \cdot H_2O$) appear in two forms, of which boehmite is adaptable to the Bayer process and diaspore is not. Monohydrates are the common bauxite ores of Europe and may be found in varying percentages in other bauxites. They are less desirable than trihydrates ($Al_2O_3 \cdot 3H_2O$) found in most other sources in the form of gibbsite. The trihydrates are preferred because they are more soluble in caustic solution and can be handled at lower temperature and pressure in the Bayer process.

Impurities may limit the value of bauxite because they entail loss of soda and alumina in subsequent processing and because they increase the bulk of materials handled. The three most common impurities are silica and iron oxides and titanium oxides. Reactive silica is the most objectionable because it combines readily with soda and alumina to cause processing losses. Hence, low-silica content is a much sought characteristic of the ore. Typical analysis of good quality bauxites are as follows:[7]

	Trihydrate	Monohydrate
Al_2O_3	57	53
SiO_2	4	7
Fe_2O_3	5.5	25
TiO_2	1.5	3
H_2O	32	12

Nearly all bauxite mined currently is produced by open pit mining, although some underground mining is done in Europe. Open cast mining can be accomplished very cheaply under conditions where scale of operation and depth of overburden are fav-

[6] United Kingdom Overseas Geological Surveys, *op. cit.*, pp. 3-4.
[7] Reimers, ST/ECLA/Conf.11/L.24, *op. cit.*, p. 21.

orable.[8] Cost will vary somewhat, depending on the nature of the terrain, hardness of the ore, and requirements for replacement of overburden. Since the material is bulky and of low value, a bauxite operation geared to distant refineries must have access to inexpensive transportation, preferably nearby tidewater.

The ore is exposed, loosened by blasting if necessary, moved by power equipment, and possibly beneficiated before shipment. Beneficiation seeks to remove as much of the silica as possible by crushing and washing. The shipping weight of the bauxite may be further reduced by drying, which removes some of the free moisture from the ore.

Most bauxite is produced under conditions which make it a very inexpensive material.[9] It is abundant, found in a number of locations, easily mined, and shipped as a bulk material, and requires very little processing prior to shipment. The investment in a bauxite mining operation is much less than at subsequent stages of the industry.[10] Aside from the earth-moving equipment, most of which is not fixed to the site, the beneficiation facilities—crushing and drying—are the principal works along with transportation and ore handling facilities.[11] The latter obviously may vary greatly depending on the site.

PRODUCTION OF ALUMINA

With existing technology nearly all bauxite used for the production of aluminum is first subjected to the Bayer process which converts it into alumina. Alumina can be made from materials other than bauxite and by processes other than the Bayer process and aluminum can be produced by processes which bypass the production of alumina, but up to the present the conversion of bauxite into alumina by the Bayer process for subsequent reduction into aluminum is the prevalent technology.

[8] *Ibid.*, pp. 15-16.
[9] See Chapter 7, pp. 149–50.
[10] Pierre Crosson, Gregory C. Zec, and Francis J. Kelly, *Capital Coefficients for the Integrated Aluminum Industry* (Washington: U.S. Bureau of Mines, November 1953), mimeo., pp. 12-25. This study shows investment of $3.70 per ton of annual capacity in 1952. Wide variation could be expected and current figures are surely higher.
[11] *Ibid.*, pp. 12-13.

The Bayer process is used to remove the impurities and chemically combined water, leaving aluminum oxide in a nearly pure state. The first step is the mixing of powdered bauxite, caustic soda (which may be made from lime [CaO] and soda ash [Na_2CO_3] or caustic may be added in liquid form), and hot water which are digested at elevated temperature and pressure.[12] In this stage the alumina is taken into solution as sodium aluminate ($NaAlO_2$) while silica, iron oxide, titanium, and other impurities are insoluble. However, the reactive silica forms an insoluble sodium aluminum silicate, thereby consuming caustic soda and alumina in the process at the rate of approximately one part soda and one part alumina per part of reactive silica. This loss of caustic and alumina imposes an economic limit of about 7 per cent on the silica content of ores refined by the straight Bayer process.

After digestion, the insoluble residue of impurities (called red mud) is separated by settling and filtering from the sodium aluminate solution and is discarded. The solution containing sodium aluminate is pumped into precipitators where it is seeded with fine crystals of alumina hydrate and stirred and cooled, causing about 50–60 per cent of the alumina hydrate to dissociate from the soda and precipitate out as crystals (i.e., the digestion $NaOH + Al(OH)_3 = NaAlO_2 + 2H_2O$ is reversed).[13] The mixture is pumped to thickeners and then to filters where the alumina crystals are filtered out from the caustic solution. The caustic (which still contains unprecipitated alumina) is recycled for further use while the alumina hydrate is washed and calcined to drive out the chemically combined water, reducing it to anhydrous alumina.

Trihydrate bauxite is easier to process than monohydrate. It can be digested at $130°-150°$ C and at pressures of 3.5 –7 kg/cm^2 while monohydrate is treated at $180°-250°$ C and 20–50 kg/cm^2.[14] Monohydrate long was restricted to batch processing, but continuous flow techniques are now adaptable to both types. Where mixed ores are involved, concessions must be made in the direction of higher temperatures and pressures than in standard trihydrate practice.

Ores with a high silica content may be refined by a modified

[12] P. V. Faragher, *Fundamentals of Aluminum* (Pittsburgh: Aluminum Company of America, 1959), p. 12.
[13] *Ibid.*, p. 13.
[14] Reimers, ST/ECLA/Conf.11/L.24, *op. cit.*, p. 2.

Bayer process—the so-called combination lime–soda–sinter process. When treating high-silica ores, the red mud generated by the initial Bayer stage is rich in alumina and soda which have combined with the reactive silica. Part of the alumina and soda can be recovered by mixing the mud with limestone and soda ash and sintering at 1,260° C. After this the sinter is ground and leached with water which takes sodium aluminate into solution for return to the Bayer plant digesters.[15]

Bayer-process technology has been improved over the years by more efficient conservation of heat, greater recovery of soda, and the development of continuous flow instead of batch processing. The lime–soda–sinter modification of the Bayer process has been the most revolutionary change in those cases where it is indicated. Finally, the sheer increase in size of the operating units in response to vastly greater markets has permitted the most basic kind of industrial economies-to-scale through the use of larger tanks, pumps, kilns, and heating units.

The physical works for production of alumina consist of grinders, heat and steam plants, mixers, tanks and kilns, as well as liquid and solid materials-handling facilities. The principal materials required are bauxite, fuel, caustic soda, and water. The product is alumina, with red mud a large volume waste product to be disposed of.

Technical factors for good operating practice in both trihydrate and monohydrate plants are offered by Reimers.[16] Obviously the variation is considerable, depending upon the type and quality of bauxite employed and the relative prices of other inputs as well as upon the scale of the plant. He figures capital costs per ton of annual trihydrate capacity for a large plant at $110–$150 and for monohydrate at $140–$180.[17] Other operating factors that Reimers[18] suggests are as follows[19] (per metric ton of alumina):

[15] *Ibid.*, pp. 4-5.
[16] *Ibid.*, pp. 21-25.
[17] The figures are for plants of over 330,000 tons. Plants of 100,000 tons would run $50-$60 per ton higher (*ibid.*, p. 21). It is interesting to compare these figures with the 1952 estimates of Crosson *et al.*, *op. cit.*, pp. 11-17, of about $88 per ton for a plant including lime-soda-sinter works.
[18] Reimers, ST/ECLA/Conf.11/L.24, *op. cit.*, p. 22.
[19] For alternative set of materials required, see U.S. Bureau of Mines, *Raw Materials for Aluminum Production*, Information Circular 7675, by D. D. Blue (Washington: U.S. Government Printing Office, 1954). See also United Kingdom Overseas Geological Surveys, *op. cit.*, p. 51.

	Trihydrate	Monohydrate
Bauxite (ton)	2.1	2.5
Caustic (kg NaOH)	80	140
Steam (ton)	2.0	2.4
Power (kwh)	200	275
Fuel (liters of fuel oil)	130	130
Labor (man-hours)	3	4
Maintenance (dollars)	3.00	4.00

The amount of bauxite required will depend on its quality. The percentage of reactive silica will diminish the recoverable alumina by approximately an equal amount. In addition, plant losses may be 3–4 per cent. Caustic requirement varies depending upon silica content and amount of washing. Steam consumption varies with heat exchanger efficiency and, if fuel is expensive, may be reduced by added capital spending. Steam efficiency is greater for large plants—perhaps 2–3 times that for small plants. Likewise, power consumption may be 50 per cent or more higher for small plants and will be higher if hard bauxite is fed to the grinders. Labor required may be 2–3 times as great in isolated nonintegrated plants in tropical climates.[20] Requirements for the lime–soda–sinter process are quite different, involving higher capital charges, much more limestone, 2½ to 3 times the fuel, and double the labor of straight Bayer process.

Reimers suggests that there are economies-to-scale up to about 330,000 metric tons of annual capacity and that there may be further administrative economies in still larger production units involving duplicate facilities. He suggests a minimum size of 30,000–40,000 metric-ton capacity for a plant oriented to a local smelter in an underdeveloped country and 100,000–165,000 metric tons for an export plant in such countries, while self-contained units in North America would be at least 100,000–165,000 metric tons and in Europe 50,000–60,000 metric tons. In fact, all alumina plants in the United States are in the 340,000 metric-ton-and-up class, while European and Japanese plants more commonly are in the 100,000–200,000 metric-ton range.[21]

[20] Reimers, ST/ECLA/Conf.11/L.24, *op. cit.*, p. 21.
[21] U.S. Bureau of Mines, *Minerals Yearbook 1963* (Washington: U.S. Government Printing Office, 1964), p. 284; *Metal Bulletin*, "Aluminium World Survey" Special Issue (London, December 1963), pp. 127-29.

ALUMINUM REDUCTION

Production of metallic aluminum from its oxide alumina is accomplished by an electrolytic process. The basic Hall–Héroult process invented in 1886 gave birth to the industrial production of aluminum and is still used for practically all of the aluminum produced. Again, it is possible to produce aluminum by other means, and there now are real prospects that this may be done, but current technology is almost exclusively based on the electrolytic process.

In this step, metallic aluminum is separated from the oxide by passing an electric current (DC) through a molten bath containing alumina. This is accomplished in a shallow carbon-lined steel vessel containing molten cryolite (Na_3AlF_6) which serves as an electrolyte in which alumina is dissolved. A carbon anode is suspended into the electrolyte. The oxygen in the alumina combines with the carbon of the anode to form carbon dioxide and carbon monoxide, consuming the anode in the process. The metallic aluminum is released in a molten condition and, being heavier than the electrolyte, remains at the bottom of the pot where it is periodically tapped off. Reduction is a continuous process, with pots being fed and tapped as required.

Since it is uneconomical to move current at extremely low voltages and since the voltage drop per cell is only about 5V, the reduction cells are employed in series or lines, the anode of one cell being connected with the cathode of the next, and all fed from a common power source. A typical potline fed by a single set of rectifiers contains up to 150–160 cells.[22] Generally, the rectifiers are housed in a long narrow building at the head and end of the line, the first and last cell of the line being placed on or as near as possible to the rectifier building. The physical layout must provide for bulk handling of materials and equipment and for dissipation of heat and control of fumes. Thus, in tropical climates or where large cells are employed, larger building space may be required to dissipate the heat, and in settled areas where control of fumes is most important, more elaborate air-cleaning equipment is required.[23] Difficulties in controlling fumes may impose a limit on size of the plant in some circumstances.

[22] Reimers, ST/ECLA/Conf.11/L.24, *op. cit.*, p. 27.
[23] *Ibid.*, pp. 27-28.

Two types of anodes are employed: the prebaked anode and the continuous or Soderberg anode. In the prebaked system a number of prebaked carbon blocks connected to electrical bars are suspended in each cell and lowered as required until consumed up to their stubs. Soderberg anodes are made of "green" or unbaked paste fed into a casing. They are baked by the heat of the pot as they are lowered into it and are constantly renewed with new paste at the top. Both types are used in new plants, the choice depending upon the specific circumstances. This choice will affect the capital costs and operating practices, however.

Reimers lists the considerations as follows: *Prebaked systems* (1) consume less power because of greater conductivity of the carbon, (2) are easier for inexperienced crews to operate, (3) form less noxious gases, but, (4) gas is difficult to collect. *Soderberg systems* (1) save on the capital cost of a carbon baking plant, (2) save on labor in manufacturing and handling carbons—more suited to mechanization, and (3) gas is more noxious and harder to clean but is easier to collect.[24] Generally, a small plant will employ Soderberg units to avoid an uneconomically small carbon-baking facility, while larger plants commonly favor prebaked anodes.

The process of electrolysis consumes principally alumina, carbon, and electric power and, of course, requires labor and capital equipment. Additionally some electrolyte is lost in the process but this is a comparatively small item of cost. Economies may be sought in any of these areas, although in practice virtually all of the aluminum content of alumina is recovered as metal, so there is scant room for improvement there.

Operating practice and quality of materials can affect carbon consumption. Anodes generally are made of petroleum coke with a pitch binder. (Impurities must be avoided since the ash falls into the pot and could contaminate the metal.) If the content of alumina dissolved in the electrolyte falls too low, the so-called "anode effect" occurs: the voltage drop increases, gas and heat form at the anode, and rapid burning of the anode takes place.[25] Present practice generally requires 0.5–0.6 kg of electrode per kg of aluminum produced, although some range is possible. Soderberg plants use more than prebaked, 0.52–0.53 kg compared with 0.45–0.48 kg under best practice. Reimers gives U.S. carbon costs as \$55–\$70

[24] *Ibid.*, pp. 6, 35.
[25] Edwards *et al.*, *op. cit.*, Chap. 9.

per ton of carbon for prebaked anodes and $40–$50 per ton of Soderberg paste.[26]

Alumina consumption is constant at 1.9 kg per kg of metal.[27] Care must be taken to avoid spillage and to recover sweepings.

Power consumption is a function of plant design and operating practice. Under the principles of Farraday's Law, the electrolysis should require only 1.7V of the 4.8V typically lost per pot. Electrical efficiency based on this ratio is on the order of 35 per cent. Some of the power not used in electrolysis is required to maintain the molten condition of the pot which operates at about 950°C, but the rest is lost through resistance in the electrolyte, at contact surfaces, in conductors, or is dissipated in the pot as heat.[28] Greater thermal efficiency may be obtained at somewhat lower current density or by reducing the operating distance between anode and cathode, thereby reducing resistance and heat loss. However, this may lower the output of the pot and requires closer attention; therefore maximum thermal efficiency may not be sought, or the degree to which it is sought will depend on the specific relationship of costs. Lowered power consumption, however, remains one of the promising areas for further improvement. A modern pot may use 17–20 kwh per kg of metal, although some of the most efficient plants may achieve figures as low as 14 kwh per kg for electrolysis.[29] Soderberg plants use slightly more than prebaked. Since electric energy is a major input whose cost varies greatly on a geographical basis, it long has been viewed as the key locational factor for the industry.

Labor is required to tend the pots, which means to maintain the dissolved alumina content in the proper range, replace and adjust the anode position, tap off and cast molten metal, operate the carbon plant, if any, and for general housekeeping, repair, and administration. A frozen crust which forms on the top of the electrolyte must be broken when alumina is added. Anode consumption may be at the rate of 1 inch per day with a 2- to 4-inch gap from the metal pool desired. Thus the anodes require attention. Pots may be tapped 3–7 times per week. Also occasional additions to the electrolyte may be called for, although in principle the electrolyte

[26] Reimers, ST/ECLA/Conf.11/L.24, *op. cit.*, p. 31.

[27] T. G. Pearson, *The Chemical Background of the Aluminum Industry* (London: Royal Institute of Chemistry Lectures, Monographs and Reports, 1955), p. 62.

[28] *Ibid.*, p. 37.

[29] *Ibid.*, p. 63; Reimers, ST/ECLA/Conf.11/L.24, *op. cit.*, p. 30.

is neither consumed nor transformed by the process and needs only limited attention. Alloying ingredients may be charged to the pot if desired. It is possible to reduce labor requirements by utilizing larger pots, mechanizing some of the operations, and, of course, by use of skilled personnel. Reimers offers figures as low as 11–14 man-hours per metric ton in the United States and France, but finds 18 hours (prebaked) and 15 hours (Soderberg) as typical for North America, while in Europe and Japan it may be 25–30 hours. He feels that 30–50 man-hours, or even more for small plants, would be required in underdeveloped countries where mechanization would not be carried as far.[30]

As has already been indicated, capital costs may vary depending upon scale, location, skill of work force, and relative cost of other inputs. Larger plants have significantly lower unit capital costs at least up to 100,000 metric tons and perhaps to 200,000 metric tons per year. Soderberg plants require no carbon-baking facility and capital needs run somewhat lower than prebaked. Plants in tropical climates face greater heat dissipation problems and thus require more square footage of building. In densely populated areas control of noxious gases adds to capital costs. Reimers contends that capital costs for plants *of the same design* would be lower in Europe than in North America while in underdeveloped areas the cost would be substantially higher than elsewhere. However, European plants typically are smaller and require more elaborate air-cleaning facilities, so cost per ton of capacity in fact is higher than in North America. Distance from equipment sources, exigencies of climate, and need to provide more complete maintenance facilities raise cost in underdeveloped areas.[31] Moreover, plants oriented to a local market in such areas normally are quite small and accordingly of high cost per unit of capacity.

Plants may be so designed and operated as to save on power or

[30] Reimers, ST/ECLA/Conf.11/L.24, *op. cit.*, pp. 31-32. Production-worker man-hours per metric ton averaged 14.5 in the United States in 1962 (derived from U.S. Bureau of the Census, *Annual Survey of Manufactures, 1962*, p. 40). Note that Péchiney's Nogueres plant requires only 4½ man-hours of potroom labor per ton according to company sources. See also Phillipe Leurquin, *Marché Commun et Localizations* (Louvain: Editions Nauwelaerts, 1962), p. 226. Finally, Von Salmuth attributes 650 persons to the French-operated Cameroon plant or the equivalent of about 25 man-hours per ton. Curt Freiherr Von Salmuth, *Handbuch der Aluminium Wirtschaft* (Frankfurt: Agenor-Verlag, 1963), p. 30.

[31] Reimers, ST/ECLA/Conf.11/L.24, *op. cit.*, p. 28.

on labor costs. Power savings may be accomplished by larger dimension bus bars or by more elaborate bus-bar design as well as by careful anode practice under prevailing technology. Labor can be economized by increased automation, particularly in pot feeding and anode adjustment. In each case, however, these economies are available only at greater capital cost.

The trend has been toward larger pots, primarily to save on labor costs. However, pots in the 70,000- to 100,000-amp range are believed to be most economical from an investment standpoint. Beyond that, size problems of ventilation and control of electromagnetic disturbance created by high amperages act to raise cost, while current density may be diminished in an effort to control heat, thereby requiring large vessels. At present, cells range from 40,000–130,000 amps and still larger ones are being tested. However, for underdeveloped countries Reimers commends cells in the 50,000- to 80,000-amp range as economical and less beset with operating problems.[32]

Reimers gives typical capital costs for reduction plants as shown below. They exclude power generation and transmission and any townsite or harbor costs but include all power costs for transformers and step-down from arrival at plant, carbon baking, casting facilities, maintenance, laboratories, offices, etc., necessary to the smelting plant.[33]

Capacity	Dollars per Metric Ton of Annual Capacity	
(metric ton per year)	Prebake	Soderberg
20,000	1,000–1,300	900–1,200
50,000	750–1,050	700–1,000
100,000	650–850	650–850
200,000	500–700	550–750

Of the equipment required in a reduction plant, pots, bus bars, and electrical equipment are by far the largest items of expense, followed by materials-handling facilities. In addition, a substantial building is required to house the bulky equipment involved.[34]

With capital cost so important as an element of total cost, the

[32] *Ibid.*, pp. 34-35.
[33] *Ibid.*, p. 26. Crosson *et al., op. cit.*, showed a figure of $547 per annual ton of capacity as of 1952. Allowing for subsequent cost increases, this is quite consistent with the range shown by Reimers.
[34] Crosson *et al., op. cit.*, pp. 18-20, 113-16.

advantage of large plants is apparent. This advantage is reinforced by the fact that economies in administration and materials handling also can be had at large plants without any obvious disadvantages of size except those of a marketing nature.

It is often less expensive to expand existing plants than to build the same capacity at a fresh site. Material handling and transportation facilities, administrative and maintenance facilities, and frequently electrical equipment may be able to accommodate added loads. Potlines may be added at two-thirds to three-quarters the cost of new facilities and often more cells may be added to existing potlines at still less cost.[35] Not infrequently plants are designed so as to permit economical expansion.[36]

With existing technology, electrolyte is not a major item of expense, since little of it is consumed in the process. Natural or synthetic cryolite (Na_3AlF_6) is used because of its ability to take alumina into solution and because of its electrical conductivity. Natural cryolite is in short supply, but synthetic cryolite is obtainable. The price for cryolite is about $200–$220 per metric ton for natural and synthetic respectively.[37] Consumption is only 30–100 kg per ton of metal produced.

At present, reduction cells are lined with carbon paste or blocks of much the same composition as anodes. If blocks are used, they are fitted tightly and the space between them packed with carbon paste. As a layer of sludge may form between the carbon and the molten metal, thereby increasing the electrical resistance, some of the newer techniques have sought alternative conductors for the cathode, as will be discussed later.[38] Pots must be relined occasionally if they become burned or the carbon absorbs too much other material, the frequency depending upon operating practice to some degree. Such pots must be removed from the series to be relined, a typical interval being after about 1,000 days of operation.

To summarize on reduction: the capital costs are relatively high and the chief operating costs are represented by the cost of

[35] Reimers, ST/ECLA/Conf.11/L.24, *op. cit.*, p. 28.

[36] The Evansville plant of Alcoa and the New Johnsonville plant of Consolidated Aluminum are examples of this.

[37] This figure may be too high. A news report in the *American Metal Market* of July 26, 1963 recounted the sale of U.S. stockpile synthetic cryolite at $130 per ton. Reimers cites much higher figures of $300-$360 for natural and synthetic cryolite (Reimers, ST/ECLA/Conf.11/L.24, *op. cit.*, p. 19).

[38] See Chapter 8, pp. 165–67.

alumina, carbon, electric power, and labor, with lesser amounts for electrolyte and maintenance. Typical operating ratios per ton of metal as given by Reimers[39] are as follows:

	Prebake	Soderberg
Electric power (kwh)	17,000	17,500
Alumina (ton)	2	2
Fluoride for electrolyte (kg F)	2.5	3.5
Carbon (net kg)	500	560
Labor (man-hours)	18	15
Potroom only	8	7
Operating and maintenance supplies (dollars)	20	18

A NOTE ON FABRICATION

The fabrication of aluminum is done in a manner very similar to that of other metals, except that the high degree of workability of the metal favors greater use of extrusions and foil or very thin sheet. The equipment employed is standard metal-working equipment; theoretically, mills would be able to convert between aluminum and other metals with only minor modifications of their lines.

Castings consume a major share of the metal. In the United States in 1963 shipments of castings were equal to about 16 per cent of all shipments to domestic consumers.[40] The automobile industry, which is a large consumer of aluminum castings, buys some primary metal direct from the smelter in a molten form for this use. Secondary metal is a common source of aluminum for castings and it may be mixed with primary where necessary to achieve desired quality.

For other uses, ingot is converted into basic mill shapes (semifabrications) in hot rolling mills which produce plate, coiled reroll sheet, bar, and rod. In other cases the ingot may be extruded into desired shapes. The various products may be further processed into cold rolled sheet, plate, foil and circles, cold rolled rod and bar, and drawn wire and cable, in which forms the metal may be shipped to end users. In addition, some aluminum is shipped as powder.

Capital and operating requirements for fabrication plants vary enormously. A continuous hot rolling mill for sheet is expensive and

[39] Reimers, ST/ECLA/Conf.11/L.24, op. cit., p. 29.
[40] The Aluminum Association, The Aluminum Industry Annual Statistical Review (New York, 1963), pp. 17, 23.

requires large tonnage to be justified. The same is true for rod mills. However, the use of less sophisticated equipment with much lower capacity makes it feasible for a smaller plant to perform at least some range of fabrication for a domestic market. Extrusions and castings in particular can be fabricated on a small scale, and cold rolling mills commonly are of much smaller capacity than hot roll mills. As a consequence many countries that do not possess primary aluminum industries nonetheless engage in some range of fabrication using purchased reroll stock or even ingot, and a country contemplating a small-scale primary industry to serve a domestic market might also expect to fabricate the metal locally. It should be recognized, however, that for truly modern and complete facilities the investment required may match that of other stages of the industry.[41]

CHARACTERISTICS OF PREVAILING TECHNOLOGY

The foregoing has described the processes, equipment, and materials employed in the production of aluminum. Several characteristics of the production process which will affect location should be noted:

(1) *State of technology*. Existing technology of the industry is well developed and widely available. Although revolutionary changes are under consideration (see Chapter 8), at present a rather well-established and uniform technology prevails. It is fairly readily accessible to new producers and is adaptable to less-developed countries.

(2) *Scale*. At most stages aluminum is a large-scale industry for sound technical reasons. Bauxite mining need not be on a large scale, but if attendant port and rail facilities are necessary, and they often are, a sizable operation is called for. Alumina production is a large-scale continuous process where great efficiency is gained from sizable vessels, heating, and materials-handling equipment. Efficient reduction is feasible on a somewhat smaller scale than alumina production, yet economies in use of materials handling, administration, and electrical equipment and other plant

---

[41] Reimers, ST/ECLA/Conf.11/L.24, *op. cit.*, p. 37, suggests $500-$1,500 per ton of annual capacity for complete rolling-mill facilities. Crosson *et. al.*, *op. cit.*, pp. 13, 16, 20, and 25, show a 1952 cost of $638.64 per ton of annual capacity for combined bauxite, alumina, and primary facilities as compared with $885.82 for balanced rolling and drawing facilities.

favor establishments of 200,000 annual-ton capacity or more. There is much dispersion in the economical scale for various types of fabricating facilities. However, the two central steps in the aluminum industry—alumina production and primary smelting—are both best carried on at large scale.

(3) *Capital intensiveness.* Aluminum is a very capital-intensive industry. Investment for an integrated operation of efficient scale, excluding both electric power plant and fabricating facilities, may be on the order of $1,000 per ton. This is a figure far above that for an equivalent amount of steel-making capacity.[42] Even at full operation capital costs are an unusually large share of total costs. Capital intensiveness combined with advantages to scale means that large investments are required and therefore investors are impelled to seek politically secure locations for such expensive works.

(4) *Electric power.* Low-priced electric power is important to the production of aluminum at competitive costs.[43] Expensive power quickly makes a location untenable for the industry under normal circumstances of exposure to trade. However, other elements of cost are sufficiently important that a favorable juxtaposition of such other costs may make a site attractive even though substantially less expensive power could be had elsewhere.

(5) *Transportation.* The logistics of the aluminum industry are complex. Metal is consumed mostly in the large industrial centers but the raw materials required to make it are bulky and increasingly distant. In practice, transportation of raw materials, while a significant element of total cost, has not controlled the location of smelters although it has influenced the location of alumina plants. An effort to minimize transport costs to market in the face of changing material sources may affect the location of future facilities. So far, however, the economy of moving alumina as contrasted with the higher cost of shipping metal has reinforced the position of smelters located nearer to major markets.[44]

[42] One writer, for example, says aluminum capacity costs seven times as much (Leurquin, *op. cit.*, p. 226). Others have suggested figures of about three times as much.

[43] At present rates of 17 kwh per kg of metal produced, the cost per kg increases 1.7 cents for each increase of 1 mill per kwh in the cost of power. Existing spreads in cost of power used for aluminum reduction range from 1.5 to 8.0 mills or from 2.6 to 13.6 cents per kg of metal produced (Reimers, ST/ECLA/Conf.11/L.24, *op. cit.*, p. 33).

[44] See Chapter 7, pp. 155–62, for discussion of this point.

CHAPTER FIVE

STRUCTURE OF THE
INTERNATIONAL INDUSTRY[1]

A realistic account of prospective location of the aluminum industry must consider the structure of the industry, i.e., the business units that make location decisions and the varying resource, market, and policy situations which they face. Outside the Communist bloc a small handful of major North American and European firms own most of the known bauxite deposits, produce most of the world's alumina, smelt most of the metal, and fabricate a major share of that which is produced. These firms typically have great financial power and operate on an increasingly international scale. They are fully integrated, occupying strategic positions in the industry from raw materials to marketing, and they maintain extensive research, sales, and customer-service staffs. Their internal reasons for location of their operations must be taken into account by governments and competitors, and it would be most difficult to conceive of major new projects of international significance which did not require the participation of one or more of these firms.

Apart from the major international firms, there are a number of lesser producers whose operations are more restricted in scope. Such firms often are less completely integrated and the geographical reach of their operations commonly is much narrower. They may lack the organizational, technical, and financial resources to impinge seriously on the international scene and sometimes are themselves allied with or partially owned by major international firms. In other cases metal is produced in diversified firms in which alu-

[1] Information on which this section is based has been drawn from four principal sources: Curt Freiherr Von Salmuth, *Handbuch der Aluminium Wirtschaft* (Frankfurt: Agenor-Verlag, 1963); *Metal Bulletin*, "Aluminium World Survey" Special Issue (London, December 1963); U.S. Bureau of Mines, *Minerals Yearbooks;* and company annual reports, as well as discussions with company officials. Other sources are also cited.

99

minum is not the major interest. Finally, there are some state-owned firms which tend to be restricted in area of operation. Although it is not easy to shift into the category of major firms, certain lesser North American and European producers now possess the basis for such a shift if they so desire.

The major firms (and many of the lesser ones as well) generally have been the beneficiaries of state policies that strongly favored their growth.[2] In North America this has taken the form of favorable terms of acquisition of war-built plants, financial assistance, tax privileges, guaranteed markets and, in some cases, access to publicly financed power.[3] Tariff protection, albeit declining, also has strengthened the domestic position of U.S. firms. In France electricity has been available on favorable terms and the tariff traditionally has been high. The latter is also true of Switzerland. Thus, in each case the major firms have survived and grown with state support at their domestic base. In the main this has reflected the requirements of defense policy as perceived by the home government. The competitive strength of the major firms reflects this historical development, even though their future expansions may not benefit in like manner.

MAJOR INTERNATIONAL COMPANIES

There are six companies that operate on a major international scale.[4] Four are originally North American firms and all six have (or soon will have) smelting operations in North America, either wholly owned or through partnerships. After completion of announced expansion, all will be represented by smelting operations in Europe, either wholly or partially owned. The European picture is further complicated by its division into two major trading groups, with the North American firms only weakly represented at the smelting level in European Economic Community (EEC) countries. Outside of Europe and North America it is a rare local firm that does not have some connection with one of the six companies.

---

[2] Phillipe Leurquin, *Marché Commun et Localizations* (Louvain: Editions Neuwelwaerts, 1962), Chap. 1.

[3] Donald B. Macurda, *The World Outlook for Aluminum* (New York: F. S. Smithers and Co., 1962), p. 6; and Dominick and Dominick, *An Analysis of the Aluminum Industry in North America* (New York, 1962).

[4] They are Alcan, Alcoa, Reynolds, Kaiser, Péchiney, and Alusuisse.

100

These major firms owned directly some 63.9 per cent of the non-Communist-bloc smelting capacity in 1963 and had a share of ownership in an additional 12.7 per cent of the total.[5] Beyond this, they are most active in generating plans for additional capacity; one or more of them commonly will be mentioned in connection with any scheme for a new smelter outside the Communist bloc. The North Americans are the giants among the international firms, all being of similar order of magnitude in size and capability. The two European representatives, although smaller, deserve inclusion mainly because of the reach and diversity of their operations and their apparent determination to expand. It is worthwhile to look briefly at the situation of each of the major firms.

Alcan Aluminium, Ltd.

This is the international firm par excellence.[6] With its major production in the small Canadian market, Alcan perforce has been the world's pre-eminent exporter, finding its best markets in the United Kingdom and the United States. Its Canadian operations were expanded during World War II in response to the stimulus of the Allied war effort and with financial assistance from the United States and United Kingdom. Further stimulus came with the U.S. metal shortage of the Korean War period which enabled Alcan to negotiate long-term contracts for the sale of metal in the U.S. market.[7]

Canadian metal traditionally found a market in the United States among independent fabricators. It was competitively priced and considered a more reliable supply under conditions where buyers often feared that U.S. firms might divert metal in tight supply to their own fabricators. By abstaining from the fabrication end of the business in the United States, Alcan was better positioned to supply

---

[5] Derived from U.S. Bureau of Mines, *Minerals Yearbook 1963*, "Aluminum" chapter (Washington: U.S. Government Printing Office, 1964). Bachmann attributes 85 per cent of non-Soviet capacity to these firms (United Nations Conference on Trade and Development, *Aluminum as an Export Industry*, E/Conf.46/P/10, February 4, 1964, by Hans Bachmann).

[6] Often called a Canadian firm, it is worth noting that only 25 per cent of the shares are owned by Canadians, with 70 per cent in the United States. (From a speech by Nathanael V. Davis, President, Alcan, to New York Society of Security Analysts, June 9, 1964.)

[7] Dominick and Dominick, *op. cit.*, p. 32.

the independents and avoided the anomalous situation of being in competition with its own customers. The development of a surplus of metal in the United States and the resulting echoes in other markets in the late 1950's compelled Alcan to alter its tactics. U.S. firms aggressively acquired fabricators or expanded existing plants both in the United States and elsewhere as outlets for their metal, thereby threatening traditional Alcan markets. The Alcan response has been twofold. First, they have diversified their sales to innumerable small markets about the world which in total have become a significant part of their business. Second, following the lead of U.S. firms, they have moved into the fabricating business so as to secure their outlets for metal.[8]

Alcan is fully integrated, boasting diverse bauxite holdings in Guyana, Jamaica, France, Malaya, and Sarawak. Principal alumina plants are found in Quebec, Jamaica, and Guyana, with local capacity to serve affiliates in Brazil, India, Japan, and Norway. The firm also is participating in the Australian alumina plant now being built in Queensland. Alcan provides substantial amounts of alumina to Norway, taking metal in exchange. Canadian smelter capacity is mostly in Quebec, with a secondary center in British Columbia. All Canadian smelters are based on company-owned hydroelectricity. Local smelters, wholly or partially owned, are found in Brazil, India, Italy, Norway, and Japan; the Japanese affiliate is a partially owned firm which is that country's largest. According to a recent announcement, a new smelter is contemplated for Australia. At the fabricating level, Alcan has major plants in Canada, the United States, and the United Kingdom, but also occupies an important position, either directly or through partners, in Australia, Argentina, Brazil, India, Japan, Germany, Denmark, Netherlands, Norway, Sweden, Spain, and Switzerland. According to a recent announcement it plans a very large rolling mill in Germany to be owned jointly with the German state firm.

The surplus of metal in the United States in the late 1950's and early 1960's and the strong U.S. movement into the British market compelled Alcan to try to diversify its sales. Meanwhile the growth of the EEC has made access to that market more difficult, and the tendency for small countries to establish smelters to serve local

---

[8] Whereas 28 per cent of sales were in fabricated or captive form in 1956, by 1963 the figure was 56 per cent and aimed still higher (speech by Nathanael V. Davis, *op. cit.*).

markets—a trend visible in several countries—has posed a further threat to Alcan markets.

This problem was aggravated by the heavy commitment to expand capacity first at the partially built Kitimat site in Canada which meant that Alcan was necessarily less nimble about grasping local opportunities to locate smelters elsewhere in the world. A second consequence of the commitment to production in Canada is that Alcan is strongly in favor of minimal restrictions on trade in aluminum. With only a small domestic market in relation to Canadian capacity, it must press for the easiest access to other markets.

Alcan's situation has been unfavorable with respect to present maneuverability, but the company has other strength. Its cost of production is conceded to be exceptionally low and its expansion likewise can be accomplished at modest cost.[9] It enjoys an excellent technical reputation and may well be the pioneer of a significant new smelting process. Its familiarity with international markets seems unsurpassed. As growth in world demand absorbs slack potential in Canada, it regains flexibility. Meanwhile it is building a solid position as a fabricator and seller in very diverse markets and can hope to profit from their future expansion as well as from any future liberalization in trade.

Aluminum Company of America (Alcoa)

As the oldest North American firm, Alcoa traditionally has been the industry's leader in the United States, where it remains the largest firm. However, like the other U.S. firms, it is a relative newcomer to major international operations (except at the bauxite stage) and, in common with the other U.S. producers and the French firm, it enjoys the advantage of a strong domestic base for its production and sales. Within the United States, Alcoa historically has been in the forefront in pioneering new uses, on occasion going so far as to enter production of end products in order to promote consumption of metal. It also enjoys a reputation on the technical side, both in production and in technical service to customers.

[9] To a greater extent than other North American firms, Alcan has invested in its own power facilities. While initial outlays are heavy, the company gains an exceptionally favorable position as these hydro plants are amortized.

103

Alcoa has a strong bauxite position in Surinam, the United States, Jamaica, and Australia. Principal alumina plants are on the Gulf Coast, with ready access to seaborne bauxite, and in Arkansas where local ore is available. New plants recently have been constructed in Australia and Surinam to serve local smelters. Reduction plants in the United States account for most of the firm's capacity. In addition, Alcoa, alone or with partners, has built or is building capacity in Norway, Australia, Mexico, and Surinam. It is worth noting that Surinam metal is expected to have duty-free access to the EEC as Surinam will be an associated country. At the fabricating level the firm traditionally has been exceptionally strong in the United States and it has connections with major fabricating plants in the United Kingdom, Japan, Australia, and Mexico.

The older Alcoa plants in the United States are based on hydroelectricity, both purchased and owned, but post-World War II expansion has stressed thermal energy and Alcoa spokesmen are on record as favoring Ohio Valley coal-based plants for future expansion. A major rationale offered for this position is the advantage of access to markets and the availability of necessary infrastructure and service facilities.

With its U.S. production widely dispersed, fully utilized, and easily expandable, Alcoa is in position to take an unencumbered look at foreign operations. Its bauxite in the Caribbean area is amenable to sea transport while both Australian and Caribbean bauxite could be moved to nearly any desired site if first refined into alumina before shipment. Alcoa's reduction operations in Mexico and Australia seem aimed chiefly at local markets, but their plants in Surinam and Norway, combined with a growing sales organization, suggest a strong interest in the European market. However, in Europe they remain weak at the fabricating level, except in England, and it would appear that they will need to repair this weakness before they can make a major impact there.

Reynolds Metals Company

Second largest of the U.S. firms, Reynolds has become a highly diversified international operation leaning heavily toward participations abroad rather than outright ownership. It has been an aggressive seller in numerous small foreign markets, but its major interest outside of the United States is in the United Kingdom.

At the bauxite level, Reynolds draws ore from Jamaica, Guyana, and Haiti as well as from the United States Arkansas deposit. Principal alumina plants are in Arkansas and Texas, reflecting reliance upon domestic and imported ore respectively. In addition, the company has major participation in British Aluminium Company, Limited, plants in the United Kingdom and in the Fria consortium. Smelters in the United States are based both on hydro- and thermal-power sources. Reynolds operates smelting capacity abroad only via its participation in British Aluminium with capacity in the United Kingdom, Norway, and Canada. However, the Reynolds name has been most prolificly associated with numerous prospective smelters about the world; they have withdrawn from a projected Greek operation but participate in a minor way in the Ghanaian plant under construction and will have a major interest in the projected Venezuelan plant. At the fabricating level the company is strongly positioned in the United States, Canada, the United Kingdom, Belgium, Japan, Germany, and the Philippines.

Reynolds has shown a flair for pushing new and unusual applications of aluminum. Its special emphasis has been in the packaging applications which initially drew the company into the aluminum business, but it has also promoted automotive, rail, and other uses.

With its influential position in British Aluminium, Reynolds has made one of the most substantial overseas ventures of any U.S. firm, gaining thereby not only a position in the United Kingdom market, but also in Canada. Recent association in fabricating facilities in Belgium and Japan suggest a long-term interest in foreign operations beyond the British affiliations and this view is supported by the degree of activity exhibited by the company in investigating overseas smelter possibilities.

Kaiser Aluminum and Chemical Corporation

While this is only the third largest U.S. firm, it has proved to be one of the most active on an international scale. Moreover, it has been one of the boldest firms with respect to location policy, basing most of its U.S. production on thermal energy and not hesitating to undertake foreign ventures where other firms have shied away.

A comparative latecomer on the international scene and without domestic bauxite sources, Kaiser has established an extremely

105

good position in bauxite reserves, producing in Jamaica **and** Australia. Jamaica is the world's foremost source of bauxite currently and the fabulous Australian deposit at Weipa, in which Kaiser has a major participation, may well be the principal source in the future. To exploit the Australian bauxite, Kaiser and its associates are building a major alumina plant in that country. U.S. smelting operations are supplied from Gulf Coast alumina plants, while minor production in Tasmania and India supports existing smelters there. Smelter capacity in the United States is divided among three large plants on the Gulf Coast, Pacific Northwest, and Ohio Valley, drawing on gas-, hydro-, and coal-energy sources respectively.[10] In addition, Kaiser has participation in plants in Tasmania, India, and Spain and has a major project under way in Ghana. The firm's name is associated with diverse projects proposed elsewhere, some of them likely to be realized soon. At the fabricating level in the United States, Kaiser is less strongly represented than the other major firms. Abroad its chief works are in Germany, the United Kingdom, India, Australia, and Argentina.

In addition to the above smelters, Kaiser has under consideration projects in New Zealand, Japan, Curaçao, and possibly Germany. Clearly it has moved onto the international stage in a major way. Of the North American firms, Kaiser has made the most daring commitment in the EEC by building a large rolling mill in Germany and announcing plans (still doubtful) for a smelter in that country. It also is the prime mover in the sole American project to date in Africa. Thus Kaiser has moved to establish both production and fabricating facilities in the principal consumption centers and will have export metal available from Ghana and the South Pacific which can be sent to any of these markets.

Péchiney Compagnie de Produits Chimique et
Électrométallurgiques

This firm long has enjoyed the predominant position in the French national market, there being only one other producer, of much smaller size, in France. Heretofore protected by a high tariff, encouraged by government policy making power available on favorable terms, and with a tight control over domestic fabri-

---

[10] A small additional plant in the Northwest has been reactivated.

cators (exercised jointly with the other French producer), Péchiney has enjoyed a very comfortable position and could expand with the growth of the French market.[11]

New fluidity has been introduced into this situation by the formation of the EEC and by the aggressiveness of the North American firms. French tariff levels have been lowered as required by the EEC, bringing the threat of foreign competition within France. The French industry might have hoped to take advantage of the enlarged common market via new plants in associated African territories, but political turbulence so far has frustrated this ambition except for one plant in Cameroon. Meanwhile the energy situation within France and the EEC has not been such as to encourage new domestic plants. Thus Péchiney, once exclusively French-based, has branched out and entered the international arena on a significant scale.

Principal bauxite source for Péchiney has been southern France, but it shares in the Fria consortium, mining and processing bauxite in Guinea. In addition, it has interests in Australia and Greece. Alumina production is near the bauxite sources in France and Guinea, and it will share in the Greek and Australian alumina plants now under construction. Smelter production is in southern France in the Alps and Pyrenees, with a thermal-based plant also located in the latter area. In addition it has a smelter in Cameroon and a share of capacity in Spain. Proposed Greek and U.S. plants with Péchiney participation exhaust its present commitments, although the Péchiney name is mentioned in connection with several other possibilities. Fabricating is concentrated in France, with major works also in Belgium and the United States.

Although much smaller than the major North American firms, Péchiney is conceded to be among the technical leaders of the industry. Faced with the erosion of its nearly exclusive position in France and with the absence of a favorable power base for expansion there, the firm is forced to become more international in outlook if it wishes to expand. Apparently the move in that direction already is under way.

[11] French producers are very sensitive to any suggestion that they get power on favorable terms, pointing out that the government has simply allowed them to buy power at costs similar to those they faced on this power prior to nationalization of company-owner power plants.

Schweizerisches Aluminium A.G. (*Alusuisse*)

Comparable in size to Péchiney, Alusuisse is the only one that does not have a predominantly national base for its operations; instead Alusuisse has widespread interests in other countries and it appears that the Swiss character of its operations will become still further diluted over time.

Currently Alusuisse mines bauxite in France, Italy, Greece, and Sierra Leone, and has a minor share in the Fria works. Its alumina is produced in Germany, France, and Italy as well as at Fria. Alusuisse also heads a consortium that will build a major alumina plant in Australia based on the Gove bauxite deposits. Smelter capacity, in most cases wholly owned, is in Germany, Switzerland, Italy, Austria, Norway, and the United States. Most of the smelters are of modest size. Alusuisse has Europe's largest rolling mill in Germany and other fabricating capacity in Switzerland, the United States, and United Kingdom. They plan a major smelter in Holland and have been mentioned in connection with several other projects.

With smelters in both the EEC and European Free Trade Association (EFTA) countries as well as in the United States, the company has a base for expansion in the two principal markets of the non-Communist world. It is known as an exceptionally well-run firm which, despite its modest size, has been very successful in diversifying and expanding its operations. Although it has enjoyed a high tariff in the Swiss market to protect its domestic production, unlike the other international firms, the domestic market has not been its major source of business. In its foreign operations, which comprise the bulk of its production, Alusuisse could not claim the kind of tax and financing assistance that so benefited the growth of other international firms. In a period where the industry is likely to be compelled to rely more completely on commercial incentives than in the past, the demonstrated capability and aggressiveness of this organization, by its very essence an international firm, is sure to make it an important factor.

SMALLER FIRMS

While the major international firms described above account for an impressive share of the world's industry, the competitive situation and policy climate of the industry also are affected by the existence of a number of smaller firms. Their circumstances vary. Some are publicly owned, others privately. Some are restricted almost exclusively to domestic sales, others are largely exporters. Their degree of integration varies, some are only divisions of large diversified firms whose major interests lie in other fields, while others are exclusively aluminum firms. Some appear ambitious to expand, and at least one is likely to qualify as a major international firm if it succeeds in realizing announced plans. Because they often impinge on local and international markets in a somewhat different way than the major international firms, it is necessary to look briefly at their situation.

Some of the smaller firms resemble their larger competitors in controlling their own bauxite and alumina supplies, but this is by no means universal. Nearly all of the bauxite produced is mined by aluminum companies and those who do not have their own supplies therefore often must buy from competitors. Likewise, most of the alumina is produced by aluminum smelting firms, although there is a sizable exception in Germany. In view of the heavy cost of plant in the industry, firms are reluctant to be dependent on competitors or other outside sources for their raw materials, for any extended shortage can be disastrous. Nonetheless many small smelting companies have been placed in this position because they lack the organizational and financial resources to locate and develop distant deposits or the locational advantages and size to build their own alumina plants. As the industry grows in size and number of firms, and as more alternative sources of materials become available, this dependence becomes less onerous, for it is possible to stagger and diversify long-term contracts so as to avoid any sudden squeeze.

At the fabricating level the smaller firms run the gamut from those who are strictly sellers of metal to those who are essentially fabricators on such a scale as to find it profitable to have their own supply. In between are various degrees of integration at the fabricating level.

In the United States in addition to the three major firms, the existing affiliate of Alusuisse, and the affiliate of Péchiney now under construction, there are three other smelters: Ormet Corporation, Harvey Aluminum, Incorporated, and Anaconda Aluminum Company. By most standards these are sizable organizations. Ormet is integrated, with good logistics and operations of efficient scale at all stages. The two principals in the Ormet firm are diversified companies fabricating and dealing in metals on a scale that makes it attractive to have their own source of aluminum.

Anaconda is a name better known for copper than for aluminum and one long connected with the state of Montana. In this case the desire for diversification into aluminum as insurance against a competing material and the availability of hydropower in its familiar home territory seem to have motivated the aluminum production. Anaconda is not fully integrated and has had to rely upon purchased alumina with an expensive rail haul from the Gulf Coast. Since Anaconda is committed to production in Montana, its disadvantageous raw-material situation prompted it to search for substitute materials and processes involving the use of clay as raw material for an unconventional process of making alumina. The company claims success for this process and has announced it will expand production on this basis, although to everyone's surprise they intend to use Georgia clay as the raw material. They are integrated at the fabricating stage.

Harvey was originally a consumer and fabricator of aluminum that took advantage of government incentives growing out of the Korean War to enter the smelting business. By origin it was integrated at the fabricating level but lacked sources of bauxite and alumina facilities. This gap was bridged by use of alumina imported from Japan while an alumina plant has been under construction in the Virgin Islands. Meanwhile Harvey has grown more ambitious, announcing intended expansion of its smelting and fabricating facilities in the United States and entering the world market in a significant way. Harvey plans to build a major smelter and fabricating plant in Norway in connection with local interests and has negotiated for bauxite in the rich Guinea deposits once claimed by Alcan. If these various expansions are realized, the firm will have become a significant factor in the international market.

In Latin America there are only small local plants and only one of these is without connection to one of the major international firms. Moreover, all of the major smelter projects so far discussed for Latin America have involved the participation of one of the international firms. The international firms, anxious to avoid being foreclosed in small but growing national markets, have made proposals for local smelters in situations where, in the absence of trade restrictions, imported supplies might be provided more cheaply. However, the companies are alert to considerations of

110

national pride and aspiration for development and the desire of governments to economize on foreign exchange. They recognize national markets can be isolated by their governments and monopolized by local plants. Therefore, if the market warrants even a minimum plant, they are prepared to negotiate for the kind of terms which would make a plant feasible so as not to see the market pre-empted by a competitor. So far only Mexico and Brazil have local smelters of this kind, but Argentina, Venezuela, and perhaps Colombia seem likely prospects. In addition, of course, there are possibilities, either here or elsewhere in Latin America, for major production on a competitive basis.

In Europe the position of the international firms is far weaker. Historically each of the significant producing countries has had a predominant national firm and there may be other smaller ones as well. Two of these already have been described. Péchiney occupies the position of predominant firm in France, accounting for about 85 per cent of French capacity, but it has significant interests abroad. Alusuisse also has a high share of Swiss capacity but is still more active abroad. In other cases the national firms tend to be more restricted in scope, even when selling abroad.

Norway is a particularly interesting case because of the diverse elements found there. The largest producer is the state firm, A/S Aardal og Sunndal Verk. This firm is unique for an operation of this size because it has no bauxite or alumina production and it does not fabricate metal. Rather it simply sells metal on the international market at a favorable price—a tactic which causes grumbling from private competitors in the world market. Compelled by the small scale of the Norwegian market to be an exporter, Aardal has succeeded in building up a clientele of independent fabricators who welcome this dependable, noncompeting, and inexpensive source of supply. It buys alumina on long-term contracts (so staggered as to give no supplier excessive leverage) paying in metal. Thus, for minimal investment and organization, and possessing the single advantage of inexpensive power, the firm has been able to compete in such export markets as the United States, United Kingdom, and Germany and it has had a significant impact thereby on world markets. It will be interesting to observe whether future expansion of this firm's production will bring integration into fabricating, possibly in overseas markets.

Other Norwegian firms have participation of a major international firm—an advantage when it comes to disposing of the metal in export markets. There is no apparent advantage to be gained from either producing alumina or fabricating metal in Norway, and the pattern of future expansion can be expected to occur largely through participation of international firms that will export metal. The Alnor project of Harvey and Norsk Hydro is a partial exception in that they propose to fabricate metal in Norway for export. The rationale for this is difficult to understand unless it was considered likely to dispose the Norwegian planning authorities more favorably toward granting the necessary concession.

Austrian production is dominated by a state firm, VMW (Vereinigte Metallwerke A.G.), which also engages in fabrication and export trade. Originally part of the German industry, it depends upon Germany for alumina and as a market for much of its output. No external expansion is likely.

Two producers account for the whole of present German output, with the predominant position again held by a state firm, VAW (Vereinigte Aluminium Werke A.G.).[12] This firm is integrated but has a relatively poor bauxite position with only a minor participation in Fria and some output in Greece. At the alumina level it is stronger, processing ore purchased from diverse sources. The company benefits from old hydropower rights and from an exceptionally favorable source of lignite. It has fabricating facilities but not enough to handle all its output.[13] With an expanding demand in Germany and a need to strengthen its raw material base, this firm could easily be tempted to extend its operations abroad. A tentative project in India is its first connection with a smelter outside of Germany. More recently VAW has acquired a fabricating plant in the United States, suggesting still further interest in foreign operations.

The major Italian producer is the diversified Montecatini group (Montecatini, Soc. Generale per l'Industria Mineraria e Chimica) which, in its aluminum operations, is fully integrated and international in its interests. As yet, however, its operations are predominantly in Italy where it enjoys the leading role as a smelter and also mines bauxite, refines alumina, and fabricates metal. At

[12] Alusuisse is the other smelter in Germany.
[13] Recent announcement of plans for a fabricating plant jointly owned with Alcan suggests a vigorous determination to defend its position in the German fabrications market.

112

present it shares the domestic market with Alusuisse at the smelter level and faces imported metal at the fabricating level. Moreover, the Italian power situation is not conducive to much domestic expansion on a commercial basis. With the lower tariffs imposed by the EEC, Montecatini will be under increased pressure in its home market and would seem a likely candidate for expansion abroad. However, this possibility may be postponed if a government policy decision makes subsidized power available to the firm in Sardinia.

Other production on a modest scale is found in Sweden and Spain. The Swedish output feeds an active local market with the single firm seemingly not more ambitious than that. In Spain most of the output involves participation of international firms, although one minor plant has mixed state and private ownership. Yugoslavia has bauxite, power, many plans, and considerable potential. However, it lacks capital and market outlets—deficiencies that are apt to limit its production unless major changes are made to welcome foreign investment. Again the focus is on the local market. Other projects under consideration elsewhere in Europe count on the participation of the major firms.

In Africa, once thought to be the promised land for aluminum production, a single plant (Cie Camerounaise de Aluminium Péchiney-Ugine [Alucam]) under Péchiney control is operating, while another under Kaiser control is being built. All other schemes are in abeyance, so far as is known.

All of the plants currently operating in India have some degree of participation by the international firms. Most proposals also involve foreign participation, although the Indians have been active in securing the interest of Communist-bloc groups and of lesser Western firms for these proposals. The Indian planning authorities aim at sharply increased production and consumption of aluminum and are able to decisively affect the market in that country via government action.

Japan has expanded output roughly apace with its very rapid growth in consumption. Four firms share in this market, the largest of them (Japan Light Metals Company, Limited) being half owned by Alcan. Again the Japanese government is in position to strongly influence the growth of domestic production not only by trade restrictions but also by means of financial controls. So far, Japanese aluminum production has grown despite the fact that Japan has no bauxite and quite expensive power. The chief ingredients would appear to be a ready market and a government policy favoring domestic production. Possible departures from this trend provide one of the more interesting subjects of speculation in the industry.

Elsewhere in non-Communist Asia there is only minor production by a state firm in Taiwan (Taiwan Aluminium Corporation). Proposals made for the Middle East, Turkey, Indonesia, and New Guinea all have involved participation of foreign interests, while any thoughts of developing major Asian rivers for large-scale aluminum production remain only a dream in the minds of planners.

In sum, then, the smaller firms play only a minor role on the international scene, though in the case of Aardal and the aspiring Harvey this is not to be ignored. In some countries of Europe and in Japan, however, they may comprise the majority of domestic production or satisfy a major share of domestic needs. In many instances these firms (except Aardal and the smaller U.S. firms) have not so far had to face the full impact of growing international competition. It seems quite likely that the major firms will strengthen their position vis-à-vis the smaller in years ahead, while a few smaller firms appear to have the capacity to broaden their operations and compete successfully.

THE STATE OF INTERNATIONAL COMPETITION

Prior to World War II the aluminum industry had a long history both of domestic monopoly and international division of markets. However, emerging from World War II, no effort has been made by the industry to restore earlier international control arrangements. In part this may be attributed to the shortage of metal and rapid growth of consumption which characterized much of the period. Under the circumstances there was little pressure on the companies to make market sharing arrangements. By the late 1950's when metal surpluses became available in some areas, the industry had so grown and the number of firms increased sufficiently that attempts at market division would have proved even more difficult than the unrewarding earlier efforts. Moreover, the climate of opinion in most industrial countries has grown more skeptical of cartel arrangements. Finally, there is keener awareness of the competition between aluminum and other metals and between primary and secondary aluminum. The industry perceives its market as expandable in response to its developmental efforts and recognizes the threat to existing markets if an effort is made to force prices up too far.

As a consequence the major international firms appear to compete strongly with each other. There is no apparent willingness to adopt a policy of leaving the other firm's market inviolate in return for like treatment. It may be necessary to defer to competitors where they are able to enlist government power to support their position in home markets, but in other cases the competition is sharp.

In the past it was customary to observe that the price of primary ingot does not fluctuate freely. Alcan, as the world's largest exporter and certainly one of the lowest cost producers, was recognized as playing a key role in stabilizing the world price within the context of a policy of balancing long-term demand and supply at a price that would attract expansion capital to the industry.[14] This comparative rigidity of ingot prices still appears in the movements of list prices, but it has become deceptive. Major suppliers are subject to an uncertain amount of competition from Soviet and Norwegian metal which is beyond their control; in periods of slack demand this apparently has contributed to some price shading. In its present volume the Russian competition has been contained by the major firms, and is not disastrous, but the majors have battled for the market by price cutting among themselves during periods of slack demand.

Competition for export markets has been keen, especially since the supply of metal became ample in the late 1950's and trade barriers of earlier postwar years were reduced. This has involved growing cross-penetration of domestic markets of major producing countries as well as exports to nonproducing and underdeveloped countries. A principal tactic has been the acquisition of captive fabricators in foreign markets which provide a continuing market for metal. A great many once-independent fabricators have been absorbed in this drive on the part of primary producers.

A second tactic, one which may successfully outflank competitors

[14] Bachmann, E/Conf.46/P.10, *op. cit.*, p. 69.

for markets in nonindustrial countries, is the construction of primary facilities in such countries behind a wall of government protection. It may then be possible for the producer to pre-empt the local market and compel captive fabricators of competitors to purchase his metal. A case in point is the recent construction of a mill in Mexico with such results, while the maneuvering among three producers to build a plant in Argentina is clearly motivated by a reluctance to see a competitor pre-empt the market. This tactic is not limited to less-developed countries. In Australia excess capacity has been built by two producers under the protection of an import quota system. As a consequence, the major fabricator, which is affiliated with a foreign producer, now is cut off from its parent source of supply and the parent firm contemplates a local smelter to supply its fabricator, thereby adding to the excess capacity. Under this kind of pressure, if governments are willing to insulate the producer from foreign competition, no market of even modest size need remain long without a primary facility, although (depending on local conditions) it may pay more dearly for its metal. Finally, it should be noted that international competition may take the form of establishing reduction facilities within major industrial country markets where the firm hopes to sell. European producers have used this tactic in the United States and announced American plans include at least one such case in Europe. The Alcan operation in Japan provides another example, and this long has been the practice of Alusuisse.

The result of these tactics is to narrow the market for independent metal. As a consequence, penetration of a market increasingly requires that a firm have the organization and funds both to produce and fabricate metal—a factor that favors the established firms.

Also of interest from the standpoint of competition is the increased popularity of international participations and consortia. If the consortium partners are essentially noncompetitive, then no reduction of competition need occur, but with the growing internationalization of the industry, the major firms find themselves in competition in an increasing number of markets. If the partners or consortium are producing for the domestic needs of a small protected market, i.e., a market where imports would be restricted in any case, then its effects on competition is negligible and may even be favorable within the host country. This is a common pattern in Japan, India, and certain nonindustrial countries. The partnerships often involve a North American or European firm together with a domestic interest and may make available to the host country the technical skill of the foreign participant.

A different situation occurs where a consortium of international firms builds productive capacity for the export market. So far these are most important at the raw-material stage with alumina ventures in Guinea, Australia, and Surinam as the most prominent examples. However, the Greek and Ghanaian schemes, oriented to export, involve international partners who in the normal course of events might be in competition with one another in some markets. A network of such arrangements, should it develop, could serve to dampen international competition. This possibility is of special importance where giant projects may come under consideration which are of too great a magnitude for the financial and marketing resources of a single firm. So far this concern is mostly hypothetical, but it may not remain so.

CHAPTER SIX

TRADE AND TRADE POLICIES

INTERNATIONAL TRADE IN ALUMINUM: GENERAL CHARACTERISTICS

As was observed in Chapter 2, aluminum has, generally speaking, been produced and consumed in Europe, North America, and Japan. Countries in other areas have produced very little primary aluminum and have not been major exporters of metal. The sole major nonindustrial exporter of aluminum is Cameroon, and this is a recent development. Among the industrial countries, however, patterns of specialization have formed and the resulting trading relationships must be noted.

Trade is mostly in the form of ingot. Some kinds of fabrication can be done on a relatively smaller scale than the production of primary metal and they require few localized resources. Also the cost of shipping fabricated products and the prevailing tariffs on them are much higher than for crude metal. In consequence, fabrication is more widely dispersed and aluminum moves in trade more frequently in raw than in fabricated form.

A few countries have become specialists in providing crude metal to an international market, most notably Canada and Norway. Their primary production far exceeds their consumption or their capacity to fabricate metal. Other countries have little or no primary capacity but have substantial fabricating industries, notably the United Kingdom and Belgium; hence they import crude metal and, in the case of Belgium in particular, re-export a significant share of it in fabricated form. Germany has significant reduction facilities but also remains a major importer of crude metal and an exporter of fabricated products. In the case of the United States and Japan, closer balance is maintained between domestic production and consumption. These various relationships imply the general pattern of trade in metal.

Barriers to trade are found in the familiar forms of tariffs, import quotas or other direct controls, and exchange controls. Crude metal is subjected to the least restrictions, since many countries lack primary facilities but will wish to protect a domestic fabricating industry. Countries which find reduction uneconomical may readily admit crude metal to supply a local fabricating industry. There is at least one case of a country with a developed primary industry, but one inadequate for its needs and not easily expanded, which protects the domestic industry behind a tariff but allows a quota of low-tariff imports to satisfy the demands of local metal users. In other cases, a quota may be used to rigidly exclude effective competition with a domestic industry, especially where the market is so small that it supports only a single domestic plant of marginally efficient scale. Finally, some developing countries, though their needs are great and their domestic production inadequate or nil, may subject imports to exchange rationing in accordance with broader objectives. The existence of these various barriers is a factor which must be considered by the major international companies in their location decisions.

Most aluminum is consumed in developed countries and this pattern will prevail over the period of concern. As was seen in Chapter 5, in such countries major international firms control both production and most fabrication. Moreover, even in nonindustrial countries, the existing fabricating facilities frequently are in the hands of major international firms.[1] Given this market structure, a major outside producer hoping to sell crude metal on the world market, i.e., chiefly in developed countries, would find a major share of his prospective market pre-empted by formal or informal ties to existing major firms. The barriers to his entry into the world market need not be insuperable, but in the absence of a liaison with a major existing firm he might need to open on a modest scale at the outset.

THE PATTERN OF TRADE

Historical data on trade in *crude* aluminum are available from Metallgesellschaft.[2] These figures, which exclude trade in semi-

[1] See, for example, the listing in *Metal Bulletin* "Aluminium World Survey" Special Issue (London, December 1963), pp. 137-51.
[2] Metallgesellschaft Aktiengesellschaft, *Metal Statistics* (Frankfurt am Main, annual).

120

manufactured metal or in products containing metal, are useful to get some idea of the longer term trend in trade. Data on net trade are shown in Table 13. Traditionally there have been very few simultaneous imports and exports of crude metal and net figures have represented the gross trade rather well, although in recent years, and notably for the United States, France, and Switzerland, this has ceased to apply. While the United States has been long an established market for foreign metal, especially Canadian, American firms recently have stepped up their own quest for export markets, thereby generating a two-way flow. France, traditionally a small exporter, has undertaken to market the Cameroon production and now has a substantial import as well as export flow. Switzerland has a more complex trade pattern resulting from the location of important Alusuisse reduction and fabrication facilities just across the German border, and for other reasons.

Of the major traders the United States generally has been an import market because of the historical relationship with Canada. An exception to this pattern occurred in 1960 because of a glut of metal in the United States and the growing range of international operations on the part of American firms. The United Kingdom likewise is a large importer and a more consistent one than the United States, since its sizable fabricating industry is supported by only minor domestic primary production. The other big net importers are Germany and Belgium.

Major net exporters have been Canada and Norway as well as Cameroon with a smaller total output.[3] On a much smaller scale, Austria also has been an exporter and the penetration by the U.S.S.R. of Western markets with growing exports in recent years has, despite the limited volume,[4] caused much consternation. Switzerland, a persistent net exporter prior to World War II, was a net importer until 1963. With the possible exception of Switzerland, none of the other major countries appears to have made a permanent shift of status between net imports and net exports of crude metal since 1925.

The major intercountry flows have been from Canada to the United States and the United Kingdom, and from Norway to a very diverse group of customers in the United Kingdom, United

[3] Figures for Cameroon are not shown in Table 13. However, the entire output as shown in Table 7 of Chapter 2 (51,500 metric tons in 1964) is exported.
[4] See Metallgesellschaft Aktiengesellschaft, *Metal Statistics, 1954-1964* (Frankfurt am Main, 1965), p. 73.

Table 13. Trade in Crude Aluminum, Selected Countries,
1948, 1955, 1960-64

(thousand metric tons)

	(+ = Net Exports; — = Net Imports)						
Area	1948	1955	1960	1961	1962	1963	1964
United States							
Imports	75.4	161.2	140.3	180.7	282.1	377.1	356.4
Exports	1.1	5.4	258.5	116.9	137.2	150.0	189.3
Net	−74.3	−155.8	+118.2	−63.8	−144.9	−227.1	−167.1
Canada							
Imports	—	0.1	0.5	0.6	3.5	1.8	3.6
Exports	296.7	463.2	500.9	441.8	522.7	576.2	569.7
Net	+296.7	+463.1	+500.4	+441.2	+519.2	+574.4	+566.1
United Kingdom							
Imports	141.3	262.6	316.3	239.7	255.2	270.6	331.6
Exports	6.6	3.7	3.8	5.6	5.5	7.6	6.8
Net	−134.7	−258.9	−311.5	−234.1	−249.7	−263.0	−342.8
Norway							
Imports	0.2	0.1	0.5	2.4	1.4	1.5	11.8
Exports	21.9	61.4	138.1	146.3	171.7	207.5	264.8
Net	+21.7	+61.3	+137.6	+143.9	+170.3	+206.0	+253.0
Germany							
Imports	16.1	43.8	181.0	138.7	118.0	120.7	165.3
Exports	—	0.6	2.7	3.8	6.0	14.2	9.8
Net	−16.1	−43.2	−178.3	−134.9	−112.0	−106.5	−155.5
France							
Imports	22.7	1.1	54.1	42.5	50.6	52.7	66.8
Exports	1.2	22.3	57.3	106.5	85.6	83.8	86.8
Net	−21.5	+21.2	+3.2	+64.0	+35.0	+31.1	+20.0
Belgium-Luxembourg							
Imports	5.5	27.1	64.4	69.6	68.3	89.1	112.6
Exports	1.9	0.9	0.9	0.6	0.8	0.8	1.1
Net	−3.6	−26.2	−63.5	−69.0	−67.5	−88.3	−111.5
Italy							
Imports	8.6	6.3	34.1	25.7	44.0	54.1	32.7
Exports	17.4	4.9	1.2	0.1	0.1	0.3	19.2
Net	+8.8	−1.4	−32.9	−25.6	−43.9	−53.8	−13.5
Austria							
Imports	—	0.1	0.3	0.1	0.8	0.7	2.0
Exports	15.1	25.4	21.7	28.3	41.8	35.4	33.9
Net	+15.1	+25.3	+21.4	+28.2	+41.0	+34.7	+31.9

Table 13 (continued)

		(+ = Net Exports; − = Net Imports)				
Area	1948	1955	1960	1961	1962	1963	1964
Sweden							
Imports	13.6	18.9	28.1	24.0	30.8	36.9	25.9
Exports	—	0.6	0.1	0.1	0.2	0.5	2.1
Net	−13.6	−18.3	−28.0	−23.9	−30.6	−36.4	−23.2
Switzerland							
Imports	7.0	9.8	16.4	10.5	11.1	4.6	6.0
Exports	10.8	4.6	7.5	6.1	8.6	18.2	19.2
Net	+3.8	−5.2	−8.9	−4.4	−2.5	+13.6	+13.2
Netherlands							
Imports	—	11.4	14.7	13.8	14.8	19.8	22.4
Exports	—	3.6	0.6	2.4	0.8	0.3	0.8
Net	—	−7.8	−14.1	−11.4	−14.0	−19.5	−21.6
Spain							
Imports	—	2.3	1.7	3.9	6.8	7.1	4.9
Exports	—	—	11.3	4.0	10.6	10.3	8.2
Net	—	−2.3	+9.6	+0.1	+3.8	+3.2	+3.3
Japan							
Imports	—	—	23.0	32.4	16.4	16.7	18.5
Exports	—	14.3	—	—	5.5	14.1	19.2
Net	—	+14.3	−23.0	−32.4	−10.9	−2.6	+0.7
U.S.S.R.							
Imports	—	7.0	—	1.0	—	—	—
Exports	—	41.6	68.0	86.0	115.7	122.1	n.a.[a]
Net	—	+34.6	+68.0	+85.0	+115.7	+122.1	n.a.

[a] Not available.

Source: Metallgesellschaft Aktiengesellschaft, *Metal Statistics* (Frankfurt am Main, annual).

States, Germany, and elsewhere. Lately the Canadians have diversified their sales, penetrating continental markets in Germany, Italy, Sweden, and Spain, and selling to numerous minor markets throughout the world, including Japan, India, Hong Kong, South Africa, and Brazil. The French import pre-eminently from Cameroon and export to Belgium, Germany, Netherlands, and the United States. Germany imports from Canada and Norway as mentioned, and also from Austria, Switzerland, France, and the United States. Italy and Scandinavia are added markets for Norwegian metal. The United States engages in considerable cross-traffic, but in general imports from Canada, France, and Norway and sells to Germany, the United Kingdom, and numerous minor markets.

The foregoing pertains only to crude metal. For the period 1948–64 we can trace the pattern of trade in all aluminum, fabricated included, for major countries. Since most trade is in ingot, the greatest net movements are similar to those sketched above for crude metal, but there is much more cross-hauling of diversified fabricated products, and certain countries which specialize in these products appear differently in totals where all products are included.

For example, Belgium, an impressive net importer of crude metal, also exports substantial amounts of fabricated metal so that its net imports on a combined basis are not so great (Table 14). To

Table 14. Trade in Aluminum, Selected Countries, 1948, 1955, 1960-64[a]

(thousand metric tons)

Area	(+ = Net Exports; — = Net Imports)						
	1948	1955	1960	1961	1962	1963	1964
United States							
Imports	80.9	181.8	178.1	230.0	340.5	419.3	408.9
Exports	49.8	20.4	279.9	147.8	187.7	212.9	264.1
Net	−31.1	−161.4	+101.8	−82.2	−152.8	−206.4	−144.8
Canada							
Imports	3.5	5.9	6.8	8.4	32.8	39.1	39.1
Exports	296.7	475.8	528.3	473.4	550.9	601.2	595.9
Net	+293.2	+469.9	+521.5	+465.0	+518.1	+562.1	+556.8
Brazil							
Imports	8.3	7.0	15.1	18.6	19.8	26.3	18.8
Exports	—	—	—	—	—	—	—
Net	−8.3	−7.0	−15.1	−18.6	−19.8	−26.3	−18.8
Mexico							
Imports	—	11.3	12.4	11.0	18.0	13.1	3.3
Exports	—	—	—	—	—	—	—
Net	—	−11.3	−12.4	−11.0	−18.0	−13.1	−3.3
Venezuela							
Imports	9.1	8.1	8.7	10.8	8.7	7.7	8.2
Exports	—	—	—	—	—	—	—
Net	−9.1	−8.1	−8.7	−10.8	−8.7	−7.7	−8.2
Colombia							
Imports	—	3.9	4.6	4.6	5.8	7.8	7.5
Exports	—	—	—	—	—	—	—
Net	—	−3.9	−4.6	−4.6	−5.8	−7.8	−7.5

Table 14 (continued)

Area	1948	1955	1960	1961	1962	1963	1964
			(+ = Net Exports; — = Net Imports)				
United Kingdom							
Imports	145.3	262.6	331.5	262.1	278.3	296.3	358.4
Exports	58.4	46.0	50.1	54.0	60.1	56.7	47.8
Net	−86.9	−216.6	−281.4	−208.1	−218.2	−239.6	−310.6
Norway							
Imports	0.9	1.9	6.8	8.7	9.3	10.3	21.7
Exports	23.1	64.2	143.5	150.7	176.2	211.7	269.5
Net	+22.2	+62.3	+136.7	+142.0	+166.9	+201.4	+247.8
Germany							
Imports	16.8	44.9	189.5	150.5	130.6	139.1	194.1
Exports	n.a.[b]	22.0	40.7	45.6	53.1	69.0	67.4
Net	—	−22.9	−148.8	−104.9	−77.5	−70.1	−126.7
France							
Imports	23.0	3.4	57.3	46.8	59.3	68.3	88.7
Exports	5.5	37.7	98.0	150.5	134.4	158.0	167.6
Net	−17.5	+34.3	+40.7	+103.7	+75.1	+89.7	+78.9
Belgium-Luxembourg							
Imports	—	31.4	74.2	82.7	84.6	105.7	131.0
Exports	—	17.0	43.8	52.1	58.6	70.2	84.9
Net	—	−14.4	−30.4	−30.6	−26.0	−35.5	−46.1
Italy							
Imports	8.7	8.5	40.2	32.4	54.4	70.4	50.1
Exports	17.9	8.1	6.7	6.9	7.4	8.6	36.6
Net	+9.2	−0.4	−33.5	−25.5	−47.0	−61.8	−13.5
Austria							
Imports	0.6	0.1	—	1.6	2.7	3.4	6.5
Exports	15.1	30.6	35.1	42.9	57.8	54.9	56.0
Net	+14.5	+30.5	+35.1	+41.3	+55.1	+51.5	+49.5
Sweden							
Imports	18.3	25.9	43.6	37.4	45.0	51.7	45.4
Exports	—	3.5	6.5	6.0	7.5	10.1	11.6
Net	−18.3	−22.4	−37.1	−31.4	−37.5	−41.6	−33.8
Netherlands							
Imports	—	21.8	32.6	31.6	32.5	41.6	52.1
Exports	—	7.4	9.8	12.7	11.6	12.2	14.5
Net	—	−14.4	−22.8	−18.9	−20.9	−29.4	−37.6

Table 14 (continued)

Area	1948	1955	1960	1961	1962	1963	1964
			(+ = Net Exports; − = Net Imports)				
Switzerland							
Imports	7.7	10.5	21.4	17.6	18.9	13.4	14.0
Exports	17.0	11.4	24.6	21.2	23.8	32.6	34.9
Net	+9.3	+0.9	+3.2	+3.6	+4.9	+19.2	+20.9
Spain							
Imports	2.2	2.8	2.3	5.6	9.8	16.3	16.5
Exports	—	—	—	8.3	14.2	14.4	11.8
Net	−2.2	−2.8	−2.3	+2.7	+4.4	−1.9	−4.7
Yugoslavia							
Imports	1.1	1.4	16.1	11.2	11.3	7.8	17.0
Exports	0.1	3.6	14.5	11.6	12.4	14.0	18.2
Net	−1.0	+2.2	−1.6	+0.4	+1.1	+6.2	+1.2
Finland							
Imports	—	9.7	17.0	16.7	18.2	15.1	19.7
Exports	—	0.2	—	—	0.6	1.1	1.2
Net	—	−9.5	−17.0	−16.7	−17.6	−14.0	−18.5
Denmark							
Imports	n.a.	8.1	16.1	14.5	16.8	17.4	22.2
Exports	n.a.	1.0	2.8	2.2	2.2	2.8	2.6
Net	n.a.	−7.1	−13.3	−12.3	−14.6	−14.6	−19.6
Iceland							
Imports	—	3.7	5.8	6.2	8.1	8.8	12.9
Exports	—	—	—	1.1	0.9	1.4	4.4
Net	—	−3.7	−5.8	−5.1	−7.2	−7.4	−8.5
Greece							
Imports	—	1.7	5.1	6.1	7.0	6.8	9.9
Exports	—	0.2	—	0.3	0.8	0.9	1.0
Net	—	−1.5	−5.1	−5.8	−6.2	−5.9	−8.9
Portugal							
Imports	—	2.6	4.6	5.7	5.2	6.0	7.1
Exports	—	—	—	—	—	—	—
Net	—	−2.6	−4.6	−5.7	−5.2	−6.0	−7.1
India							
Imports	9.6	16.1	25.4	25.6	38.9	n.a.	22.9
Exports	—	—	—	—	—	7.4	0.9
Net	−9.6	−16.1	−25.4	−25.6	−38.9	n.a.	−22.0
Japan							
Imports	—	0.03	23.5	33.2	18.4	24.1	36.2
Exports	—	24.9	10.4	10.6	18.3	32.3	36.7
Net	—	+24.9	−13.1	−22.6	−0.1	+8.2	+0.5

Table 14 (continued)

Area	1948	1955	1960	1961	1962	1963	1964
			(+ = Net Exports; — = Net Imports)				
Hong Kong							
Imports	—	3.4	19.3	12.2	8.7	9.8	15.9
Exports	—	0.2	11.8	6.9	3.6	2.7	4.2
Net	—	—3.2	—7.5	—5.3	—5.1	—7.1	—11.7
Indonesia							
Imports	—	3.9	3.0	6.4	4.2	n.a.	n.a.
Exports	—	—	—	—	—	—	—
Net	—	—3.9	—3.0	—6.4	—4.2	n.a.	n.a.
Federation of Malaysia							
Imports	—	2.7	2.0	2.5	4.5	3.4	n.a.
Exports	—	0.3	—	—	—	—	—
Net	—	—2.4	—2.0	—2.5	—4.5	—3.4	n.a.
Turkey							
Imports	—	1.2	1.7	3.0	4.2	3.9	3.2
Exports	—	—	—	—	—	—	—
Net	—	—1.2	—1.7	—3.0	—4.2	—3.9	—3.2
Thailand							
Imports	—	2.0	3.1	3.4	4.0	4.8	5.6
Exports	—	—	—	—	—	—	—
Net	—	—2.0	—3.1	—3.4	—4.0	—4.8	—5.6
Cameroon							
Imports	—	—	0.7	0.5	1.0	1.5	2.2
Exports	—	—	42.1	46.1	50.9	52.3	48.7
Net	—	—	+41.4	+45.6	+49.9	+50.8	+46.5
Nigeria							
Imports	—	1.9	3.9	4.9	5.3	6.0	6.4
Exports	—	—	—	—	—	—	—
Net	—	—1.9	—3.9	—4.9	—5.3	—6.0	—6.4
Ghana							
Imports	—	1.2	1.9	4.8	4.3	7.5	5.8
Exports	—	—	—	—	—	—	—
Net	—	—1.2	—1.9	—4.8	—4.3	—7.5	—5.8
New Zealand							
Imports	—	3.4	5.6	6.3	8.0	9.5	11.7
Exports	—	—	—	—	—	—	—
Net	—	—3.4	—5.6	—6.3	—8.0	—9.5	—11.7

[a] Differs from Table 13 in that all identifiable aluminum, including secondary and semi-fabricated metal, is included.

[b] Not available.

Source: United Nations, *Yearbook of International Trade Statistics.*

a lesser extent the same is true of Germany and the United Kingdom which, on the combined basis, become somewhat smaller net importers. When only crude metal is considered, France has appeared as a small net exporter for most recent years, but is much more impressive as an exporter of all aluminum products because much of its exports are in fabricated form. Austrian exports also are more impressive on this basis, as the Austrians fabricate much of their domestically produced primary prior to shipment. Thus, by this measure, a number of developed countries show stronger net trading positions because they ship fabricated products not merely to each other but to nonindustrial countries as well, while buying very little metal from such nonindustrial countries. In addition, certain industrial countries, especially in Europe, are net importers of both crude and fabricated metal, drawing most of their fabricated imports from their neighbors who have specialized in fabrication. The net trading positions of Belgium and the United Kingdom, neither major primary producers, benefit from this trade in fabricated products while markets are provided by such areas as the Netherlands and Sweden. The United States has large imports and exports of both crude and fabricated metal, but for 1963–64 its net position was somewhat improved by use of the more inclusive trading concept.

Among nonindustrial countries only one country—Cameroon— is a significant exporter (of crude metal). Hong Kong engages in an active export and import trade, although on a net basis, of course, it remains an importer of (Canadian) metal. Other major net importers among the nonindustrial countries and their principal suppliers as of 1964 were: Argentina (from the United States, Canada, and France), Brazil (from Canada and the United States), India (from Canada), South Africa (from Canada), and Venezuela (from the United States, Belgium, and the United Kingdom).[5] Australia and Mexico have recently added primary production capacity so as to become approximately self-sufficient and proposals for new production facilities have been advanced for each of the other countries except Hong Kong.

On a regional basis a comparison of the total consumption and production figures (including secondary), developed from Tables

[5] Figures from U.S. Department of Commerce, Business and Defense Services Administration, *Aluminum Fact Book*, Tables.

3 and 8, shows that Europe was the major deficit area with 35 per cent of all consumption in the non-Communist world and over 25 per cent of production for 1964. On the same basis the Americas were the surplus area with 56 per cent of consumption and over 65 per cent of production. These relations underlie the trading patterns discerned—the principal interregional flow being from Canada to Europe and particularly to the United Kingdom.

Aluminum is widely traded but by no means freely traded. In an atmosphere that promised to remain free of tariffs, quotas, and exchange controls, it is likely that the pattern of the industry would be quite different. In fact, however, a welter of such impediments to trade exist, and it is almost impossible to discern either the extent of their influence or their likely pattern of change.

Looking first at tariffs, the levies on ingot may be nil or low in industrialized countries which are not major producers; the United Kingdom rate, for example, is zero. Rates in major producing countries generally are higher, ranging from the *ad valorem* equivalent of a bit over 5 per cent for the United States and Canada to 9–15 per cent for Germany and Italy.[6] In less industrialized countries the rates may be very high where the objective is to protect a small-scale domestic industry as in Brazil (50 per cent) or India (35 per cent) (Table 15). The motivation is less apparent in some instances where there is no domestic industry, yet rates are high. In these cases the purpose may be to discourage imports for reasons of exchange conservation.

The most significant prospective change in tariff rates is the movement of the EEC rate toward its agreed level of 9 per cent, which will leave it substantially above the American rate. In addition, there is the unanswered question of where rates will be left by the Kennedy Round negotiations—assuming that they are successful. A 50 per cent reduction of both EEC and U.S. rates would

---

[6] Germany enjoys a reduced tariff of 5 per cent on certain amounts of metal, the yearly quota being set in light of the domestic demand. In 1965 it amounted to 80,000 tons. In addition, Austrian metal exchanged for German alumina is admitted duty-free.

Table 15. Tariffs on Certain Aluminum Products,
Selected Countries, 1964

	Rate[a]	
Countries	Ingot (per cent)	Sheets and Plate
Argentina	0	40+100%[b]
Australia	7.5	30%
Belgium	5.4	9.6%
Brazil	50	50%
Canada	5.1[c]	3 cents per lb
Colombia	5	20%
Denmark	0	7%
France	10.2	15.2%
India	35+10[b]	50+10%[b]
Indonesia	50	50–100%
Italy	15.4	18%
Japan	13	20%
Netherlands	5.4	9.6%
New Zealand	0	30–35%
Norway	0	6%
Pakistan	12.5	30%
Philippines	0	22.5%
Spain	16	17–20%
Sweden	0	3%
United Kingdom	0	12.5%
United States	5.1[c]	2.5 cents per lb
West Germany	9	14%

[a] Quoted rates ignore preferential treatment accorded trading-bloc members and take no account of turnover taxes which may be imposed.
[b] The second figure refers to a surcharge.
[c] At price of 24½ cents per lb.
Source: Kaiser Aluminum and Chemical Corporation brief to the U.S. Trade Information Committee, February 25, 1964 (revised according to subsequently available information).

leave the scales in both areas under 5 per cent—hardly a major barrier to trade between the two largest markets.[7]

Tariff rates on semi-fabricated products are generally higher by a substantial margin than those on ingot, though such rates may

[7] Press reports indicate that the EEC bargaining position will start with aluminum on the exception list for which no tariff concessions will be proposed. However, there is widespread feeling in the industry that an ultimate settlement at around 7 per cent may be reached. Of course, in the horse-trading atmosphere of a tariff negotiation, it is impossible to predict the outcome of a specific levy.

130

vary to reflect the structure of local fabricating facilities. For example, a country with capacity to roll sheet and plate but lacking wire and cable facilities might maintain a high tariff on the former but a lower tariff on the latter.

Analogous observations can be made concerning raw materials. Although the United States formally maintains tariffs on bauxite and alumina, the application of these duties has long been suspended—a clear reflection of the changed materials-supply base of the American industry. Thus, as the bauxite imports from the Caribbean area holdings of American firms mounted sharply in the early 1950's, the duty on bauxite was suspended July 16, 1954, and has not been resumed.[8] However, the ¼-cent-per-pound rate on alumina remained in force until July 17, 1957.[9] It is noteworthy that Harvey contracted the same year for Japanese alumina and since then American producers have turned more actively to location of alumina facilities outside the United States.[10]

Among the major industrial countries, tariffs are only partially effective as barriers to trade in ingot. They remain somewhat more significant in the case of fabricated products. On the whole, however, the significance of tariffs in shaping trade in aluminum and its raw materials appears to be diminishing. This is especially true with regard to the United States market and will become so for the EEC if the Kennedy Round is successful. The result may be somewhat freer exchange of ingot among leading industrial countries in the future. For many less-industrialized countries, tariffs remain as important barriers to trade in ingot as well as fabricated products. Moreover, particularly for such countries, tariffs constitute only one element of a multifaceted defense.

Other distortions of free trade stem from preferential tariff groupings such as EEC, EFTA, LAFTA (Latin American Free Trade Association), and Commonwealth preference, discriminatory export incentives, tariff quotas, import quotas, licensing, and exchange controls. An example of how some of these may work is provided by Australia, which has sought to protect a growing domestic primary industry. Although there is a tariff with preference given to the United Kingdom, the real control is exercised

---

[8] U.S. Bureau of Mines, *Minerals Yearbook, 1954* (Washington: U.S. Government Printing Office, 1958), Vol. I, p. 210.

[9] *Ibid.*, 1956, p. 242.

[10] *Ibid.*, 1956, p. 237.

by the device of import quotas. Australia determines likely consumption and domestic production for each period and fixes the proportion that users may import. Initially the licensing procedure allowed substantial imports, but as domestic production outpaced consumption, the import licenses were reduced to nil to insure that domestic production be used first.[11]

An interesting by-product of this situation is that a large local fabricator in Australia affiliated with a major international producer has been compelled to purchase the domestic ingot of competitors. The desire to avoid being "shut out" of small but promising markets in this manner may lead to a race on the part of primary producers to build primary capacity prematurely or on a small scale to serve markets more economically supplied from abroad.

The practice of excluding imports competitive with local production is widespread in Latin America. In some Latin countries premium rates are charged for exchange to pay for imports. In Brazil, for example, a prior deposit of 180 per cent of the price has been required, the excess 80 per cent being held for 5 months without interest or converted into 1-year bills at below market rates.[12]

European and U.S. devices generally are less extreme. Europe is deficient in both primary and scrap and must be prepared to import substantial amounts of metal. Nonetheless, in addition to the duties previously mentioned, assorted fees and turnover and transmission taxes are levied on imports in many instances. Japan uses administrative controls and a fairly high tariff to keep imports within desired limits.

Yet, even a thorough study of tariff and nontariff barriers to trade and their shifting pattern might not in itself give a meaningful view of the true situation facing firms in the industry. Governmental attitudes toward imports, whether formulated in response to pressures from domestic industry or for more general reasons of state, are sensed by outside firms. In view of the heavy expense of developing a market and perhaps building production facilities to serve it, few firms are willing to undertake such commitments unless they feel that the long-run climate will be favorable. Thus, the mere likelihood of the subsequent erection of trade barriers

[11] *Metal Bulletin*, July 2, 1963 (London), p. 18.
[12] Information supplied by U.S. Department of Commerce.

adverse to their interests if they should try to penetrate a market may be sufficient to dissuade the attempt.

COMPANY AND NATIONAL ATTITUDES TOWARD TRADE

The interplay between the policy objectives of the various firms and of national governments is of great significance to location and trade. Prior to World War II the industry in Europe and America was characterized by monopolies often seeking state support for the maintenance of domestic production and by national governments justifying such support on the grounds of defense posture. A special place was allotted to Canadian production in the United States and the United Kingdom. Markets outside the industrial countries were insignificant and resources in Europe and America permitted the expansion of output, while colonial areas had not yet achieved a voice of their own. The multiplication of firms has brought "disorder" to this situation, and the diminished status of aluminum as a defense material, the increased reliance on foreign raw materials, the general liberalization of trade, and the newly voiced aspiration of former colonial areas all have imparted movement to this policy climate.

In earlier years aluminum was a "metal of war" and its production was fostered by governments concerned with their industrial base.[13] This was most evident prior to World War II when national industries were built or sharply expanded in Germany, Japan, U.S.S.R., and France. In the case of France and Germany the traditional interrelationships at the raw materials level were broken prior to the war as the Germans sought expansion and the French grew fearful of the consequences. The wartime expansion in the United States and Canada permanently altered the shape of the American industry, greatly expanding capacity under favorable financial terms, increasing the number of producers, and pushing the industry to a scale where it had to rely predominantly on foreign raw materials.

Production in Europe and Japan was left at a very low level at the end of the war while United States and Canadian output attained new highs. Restoration of balance was delayed further

[13] Phillipe Leurquin, *Marché Commun et Localizations* (Louvain: Editions Nauwelaerts, 1962), Chap. 1.

when the United States encouraged expansion in the aftermath of the Korean War, but in the meantime production in Europe and Japan was advancing. With commercial demand strong and the United States engaged in building metal stockpiles as an aspect of mobilization planning, there appeared to be room for everyone. The attention of the major firms was devoted to securing their long-term supplies of bauxite, and optimism prevailed concerning the future growth of the industry. Tariffs on bauxite were suspended in the United States in adjustment to the altered source of material, and later the duties on alumina also were suspended. But as growth in the U.S. economy slackened and the new capacity inspired by U.S. policy came into production along with the expansion in other centers, the world faced a surfeit of aluminum. This occurred at the same time as the newly formed EEC began the reduction of internal tariffs and at a time when most exchange controls had been dismantled. Throughout, of course, there has been the interest of independent fabricators in cheap and reliable sources of supply, and whenever they can exert influence it will be in favor of expanded access to foreign metal. Thus, the ingredients of the new competitive situation were brought together.

For Canada the policy implications have been clear enough. With an industry oriented to export and able to produce at low cost, the Canadians have supported liberalization of trade in aluminum. At the same time they have resisted the intrusion of U.S.-fabricated products into Canada by means of anti-dumping regulations. This policy is compatible with the interests of the major Canadian producer, Alcan. It is especially suited to their interest in penetrating industrial country markets in Europe and the United States. However, Alcan produces in other markets about the world and is not always so favorably situated as in Canada. As Alcan completes its Canadian facilities, diversifies elsewhere, and moves further in the direction of fabricating, it would not be surprising if it departed from its Canadian logic. It presently operates in protected local markets in various parts of the world.

The American firms are torn between their new international status and their desire to protect the domestic industry. In the case of raw materials this issue seems resolved in favor of free trade, since the United States is deficient in bauxite and alumina is best produced nearer to the mine. Although the U.S. firms remain sensitive to foreign competition in the U.S. market, at the same time they are anxious to be able to locate plants and move metal as they choose on a world scale. As their holdings abroad grow, this flexibility becomes more important to them, and the industry moves more in the direction of free trade. This shift is illustrated by the change in position of the industry from one favoring essentially import quotas in 1961 to one favoring reciprocal reduction of tariffs to zero or the lowest possible rate in 1964.[14] Apparently the U.S. industry is convinced that it can hold its own against foreign metal in domestic markets but nonetheless would like the option of producing abroad for the U.S. market. Most importantly, however, they want maximum flexibility in their international operations, including the right to penetrate the European market with export metal from whatever source. Exposure to increased Canadian competition is one cost of such a policy, but in the changed circumstances the Canadian pressure could be partially diverted.

The United Kingdom is anxious to get its metal at low cost and seemingly has no aspirations to produce more of its own. Competition in this country is felt in the semis market, and the concern is that foreign semis producers, both integrated and others, may have cheaper sources of ingot supply. With fabrication to a large extent captive of North American smelters, the industry voice in the United Kingdom is more sensitive to potential competition in semis from the Continent and competition from Russian ingot than with the status of domestic smelting. Since the North American firms supply their fabricators in the United Kingdom from outside, they must support the free passage of ingot, but protection in the fabricated market is wanted. Naturally, Russian ingot is considered suspect as a political attack on the market and presumptively is regarded as being "dumped." Therefore, pressure is maintained to restrict the volume of this metal by government controls.

Norway, as an exporter, seeks free access to other markets. Access is formally free in EFTA countries, but Austria and Switzerland as competing producers are not good prospects, and in the United Kingdom most of the fabricating industry is captive to other

---

[14] The Aluminum Association, *Increasing Free World Aluminum Consumption*, an Aluminum Industry Report (New York, August 1961), and Kaiser Aluminum and Chemical Corporation brief to the U.S. Trade Information Committee, Washington, D.C., February 25, 1964.

producers. Sweden recently has expanded to meet more of its own needs. Thus, Norway must seek out distant markets in the United States or climb the tariff barrier to penetrate its logical market in the EEC. Some of the expansion in Norway by outside firms was undertaken on the assumption that the country would be admitted to the EEC, and this hope appears to be a factor in future plans. There is speculation that Norway might find it useful to join the EEC in future years even without the United Kingdom. Although it is only one factor of the many that would bear on such a decision, certainly the expansion of their aluminum industry would militate in this direction. The North American firms investing in Norwegian production would benefit greatly if this should occur and EEC firms might seek Norwegian connections for their own protection.

Within the EEC the national positions vary. Heretofore Benelux countries have been importers and transformers of metal and therefore interested in the cheapest sources of supply and the readiest access to markets for their fabricated products. Belgium in particular was a market for low-priced Russian metal which it worked and sold in continental markets. This Belgian interest may be somewhat muted by the acquisition of major interests in Belgian fabricating plants by the international firms, especially Péchiney. The Netherlands, traditionally a trading nation, is shortly to have its own smelter. This can be expected to give the Dutch more of a proprietary interest in the future of the EEC industry.

France is presently a net exporter of metal and fabricated products and the French dispose of the Cameroon production. However, as the domestic market grows, the French will be hard put to expand economically to meet the demand because of a shortage of low-cost power resources. Government policy in France tends to be protective of industry generally, and especially of certain key industries among which aluminum apparently is included. Thus, while it is difficult to see how the industry at present can expand domestically on an economical basis, the government may make this possible through some form of grant of privilege. Should this occur, an expanded but competitively weak French industry inevitably will be on the side of protection.

The French have been very reluctant to see tariffs on aluminum reduced to present EEC targets, but with existing plant they can live with such levels. Further reductions would soon expose them to serious outside competition. The French attitude does not appear motivated so much by a desire to themselves become supplier to the deficit areas of the EEC as it is by fear that an unprotected domestic market may be overwhelmed from the outside.[15] Unstable conditions in French Africa and commitment to a national program of nuclear power development, not yet as near to commercial application as American plants, presently deny the French industry two eventual alternatives. Thus, the chief French producer has gone into operations abroad, seeking more hospitable environments in the United States and Greece, and may build in other sites as well. This growing internationalization of Péchiney, combined with the difficulty of expanding at home, will lead to a delicate policy problem for the French government: how to maintain an "appropriate" level of protection for the domestic industry while still allowing its major producer the necessary maneuverability for international operations. Some shifting of the French position over time towards freer trade seems quite possible.

In Italy the possibilities of economical expansion to meet growing consumption are even bleaker in the immediate future. Traditionally a high-tariff country, the Italian industry is exposed to increased outside competition by attainment of the EEC tariff level. Since low-cost power is not to be found in Italy, the country would seem to be a logical importer, with sufficient concessions made to protect existing plant. Nonetheless it seems likely that a government-sponsored regional development scheme may result in a major new plant in Sardinia to provide a market for a poor-quality local coal. Such a project would be high cost, and subsequent state policy surely would press on the side of protection for the resulting domestic capacity.

Germany presents the most interesting situation in Europe. Already a substantial importer, Germany nonetheless has a considerable domestic smelting industry. Domestic production cannot be economically expanded except on the assumption that nuclear power will soon be feasible. Major North American firms have acquired fabricators in Germany in hopes of supplying the German

---

[15] In France, as elsewhere in Europe, the enormous financial power of the American firms is stressed. They note that U.S. firms can take murderous losses in order to penetrate a market and that once in they have a base for dumping on a marginal cost basis in event of low demand in the United States.

deficit, but Germany also has a comparatively high share of independent fabricators. Thus, there is potential conflict between domestic producers (including the state firm) and the independent and foreign-owned fabricators—the former seeking to defend their position and the latter interested in low duties. The government has reconciled these competing interests by use of a tariff quota allowed by the EEC which permits needed metal to enter in desired quantities at reduced rates, but provides protection beyond that level. This device requires EEC assent and cannot be extended too far. These conflicting interests, together with the bias of the postwar German government for trade liberalization, prevent Germany from adopting a strong protectionist position but likewise do not lead to a forthright free-trade position.[16]

In Japan the aluminum industry is divided among several producers, each intent on maintaining its position and prepared to match the expansion of its rivals. From an economic standpoint, expansion in Japan may seem illogical, for power is expensive and, until nuclear energy is competitive, they will have to rely largely on oil. However, the Japanese retain strong elements of autarchical feeling with regard to basic industry, even though the military rationale for this has disappeared. In this instance there seems little internal conflict between government and branches of the industry; all appear prepared to pay at least a modest cost penalty to maintain domestic control. Since Japan is now becoming a member of the Organisation of Economic Co-operation and Development (OECD) and of General Agreement on Tariff and Trade (GATT), this position will be more difficult to maintain, but the Japanese government and industry are ingenious in devising measures to preserve the desired pattern. Formal liberalization of trade may occur, while in practice the Japanese maintain tight control. They do appear reconciled to the purchase of some quantity of Siberian metal but will resist large-scale imports. They may be prepared to engage in some production abroad under Japanese aegis but on such terms as will be unlikely to open Japan to the world market.

---

[16] Recent announcement of plans for a huge fabricating plant in Germany in which VAW and Alcan will participate favors a greater German interest in trade liberalization. Moreover, German primary producers have refrained from uneconomic expansion, perhaps in recognition of the broader government policy of trade liberalization.

Indian planning appears to aim at fostering the production and consumption of aluminum as a substitute for imported copper. The objective is both foreign-exchange savings and industrial development. The pace of development is affected by availability of foreign capital and judgment on domestic priorities. Likewise policy on trade will be influenced by availability of funds and the extent of the metal deficit, with the government controlling imports. In the absence of a true market, state policy clearly will play the most important role in determining the extent to which the demand for aluminum is met and whether by imports or domestic production. The bias appears to favor domestic self-sufficiency to the extent possible.

Before American firms built smelters there only a few years ago, Australia was an undeveloped market for aluminum. Despite a sharp growth in consumption, excess capacity was created, and the government undertook to protect its domestic industry from outside competition. Presumably this protection was to be temporary in nature, but it is by no means certain that it will prove so. Beyond this, the Australian government appears ambitious to use its huge bauxite resources as the basis of an export industry in metal, but so far has been frustrated by the fact that power costs are too high to make this attractive. It may be assumed, however, that the government will pursue this policy and may seek means to make export smelters attractive. In view of the distance from European and American markets, some form of refining is logical, but it may not go beyond the alumina stage.

Elsewhere in the world only minor production is found. The same interest in further processing of raw products which has been pursued by Australia is to be found in other bauxite-producing countries. There is a limit, however, on the degree of leverage which they can exert in pursuit of this aim because bauxite is quite widely dispersed. Jamaica, Guyana, Surinam, and Guinea, in addition to Australia, all are major bauxite sources and all have succeeded in acquiring alumina plants as well. So far, only Surinam has a smelter under construction. Among less-developed countries, Cameroon, Mexico, Brazil, and Taiwan have small smelters, with capacity being built in Ghana also. Smelters in Mexico and Brazil are aimed at domestic markets and enjoy government protection. Proposals for other countries, especially in Latin America, commonly assume this kind of protection.

139

On the other hand, if a less-developed country wishes to export metal it must either be a member of a major trading group or advocate free trade in metal. Venezuela provides an interesting case. Very probably a small local smelter will be erected in Venezuela to meet local needs; in this case it will expect protection. However, Venezuela may have the potential for a much vaster output and could become an exporter to LAFTA countries. Should this materialize, the emphasis would switch from protection of an infant industry to advocacy of widest trade within LAFTA, although they might still seek barriers against metal from outside the area.

For potential exporters from less-developed countries, the obstacles to production are not merely trading restrictions but also the problems of getting capital, or of providing an acceptable investment climate, and of securing marketing channels in industrialized countries. For discussion of these problems see Chapter 10.

A NOTE ON THE COMMUNIST-BLOC INDUSTRY

While this study focusses on a world exclusive of the Communist bloc, it is not possible to ignore that grouping entirely because actual or potential trade with the bloc must be taken into account. Moreover, the cohesiveness of the two major groupings has tended to break down and inevitably this will bring about diverse forms of economic relations between countries on both sides. While this must be conceded, it is far more difficult to deal with the question of how these relations will evolve because comparatively little is known about the prospects for aluminum production and consumption in Communist countries or of their strategy concerning trade.

Western aluminum producers, aware that Soviet metal has claimed a place in world markets and mindful of the Soviet plans to harness the vast hydroelectric potential to be found in Siberia, have expressed concern that large-scale Soviet exports may be thrust onto the world market.[17] The lack of information concerning

---

[17] Soviet exports of crude metal in 1964 are listed at 122,100 metric tons. Of this amount 39,400 metric tons went to countries outside the Communist bloc. Largest Western buyers were the United Kingdom and Japan, while within the bloc East Germany was the major customer. Metallgesellschaft Aktiengesellschaft, *op. cit.*, p. 73.

Soviet costs and intentions, as well as the recognition that normal commercial considerations may not motivate the Soviet action, aggravate this concern. No one (perhaps not even the Russians) can say with assurance what they will do, but certain elements of the decision can be delineated.

At the simple level of figures on current capacity, output, and consumption there is considerable uncertainty. Only estimates of production and consumption are available for the U.S.S.R., China, and most East European countries, and these estimates do not include secondary metal. Metallgesellschaft estimates total Soviet sphere primary output at 1,285,000 metric tons in 1964 and consumption of only 65,000 metric tons less, implying a net export potential of the bloc of only a small amount. By these figures the Russians are credited with 1,000,000 metric tons of production and 850,000 metric tons of consumption, with much of the difference clearly going to other East European countries.

It is interesting to compare these figures with estimates of Soviet primary capacity. Figures obtained from the U.S. Bureau of Mines in early 1966 show Soviet capacity at 1,360,000 metric tons. The *Metal Bulletin* credited them with 1,085,000 metric tons as of 1962.[18] Von Salmuth was vague, listing plans along with operating facilities but with a general note mentioning that not all are at full operation.[19] Both *Metal Bulletin* and Von Salmuth are in general agreement that the Soviet industry aims at a capacity of 1,900,000–2,000,000 metric tons as of 1967 or thereabouts. Most of the recent and anticipated increase in capacity is to be completed in the Irkutsk and Krasnoyarsk areas deep in Siberia.

If these projects are realized, the U.S.S.R. will have available about 8 kg of primary metal per capita by the late 1960's—a figure in excess of that of most Western countries at present. While some of this could be shipped to Eastern Europe, these countries also aim at increasing their output.[20]

---

[18] U.S. Bureau of Mines, *Minerals Yearbook, 1963* (Washington: U.S. Government Printing Office, 1964), p. 225 and *Metal Bulletin,* "Aluminium World Survey" Special Issue (London, December 1963), p. 132.

[19] Curt Freiherr Von Salmuth, *Handbuch der Aluminium Wirtschaft* (Frankfurt: Agenor-Verlag, 1963), pp. 326–27.

[20] It is not expected that other East European countries will become major exporters of aluminum to the West. Neither power nor bauxite resources are available to them under conditions that would give them a competitive advantage.

Little is known of the pattern of consumption in Russia and still less of how this may alter. The original impetus for expansion was related to military needs, but Soviet military strategy, like the American, no longer requires vast numbers of military aircraft. Moreover, as a land power operating from interior lines, it needs to place less emphasis on airlift with its accompanying premium on military use of aluminum. So far as is known, aluminum has not been used extensively in construction, a major use in the West. Obvious applications are available in electrical industries and in certain types of transportation, but the consumer-oriented industries such as autos, appliances, and packaging remain undeveloped. On the other hand, Soviet authorities have paid more attention to design and style of both industrial and consumer goods in recent years, and this may be reflected in greater use of aluminum.

Nonetheless, the prospective level of output appears out of proportion to the general level of the Soviet economy and raises the specter of large-scale dumping of Russian metal in Western markets. It also raises questions concerning Russian objectives in building to this scale.

It may prove impossible to complete the expansions as quickly as hoped by Soviet planners and they may aim to apply most of the increase domestically. Certainly this amount of metal will be manageable by the early 1970's if not by the end of the current decade.[21] However, on an interim basis some surplus of metal may well develop. For the West, the long-term question of interest is whether this is simply the result of improper phasing of output and consumption or whether it represents a deliberate decision by Soviet authorities to attempt to win a larger position as an aluminum exporter.

No immediate answer to this question is apparent. It is known that Soviet authorities are subject to regional pressures, are committed to the rapid development of Siberia, and have a strong

[21] Purely as an illustration, if Soviet industrial growth averages 6 per cent per year in 1964-71 and aluminum consumption follows the Western pattern of a substantially higher growth rate, then a 9 per cent growth rate for aluminum seems plausible. Extended 1964-71, this implies Soviet consumption of 1.6 million tons by the end of the period and, if exports to Eastern Europe increase proportionately, most of the increased capacity could be absorbed by the Communist bloc. Metallgesellschaft (op. cit., p. 73) figures suggest that Soviet consumption of primary increased an average of 7.7 per cent per year 1960-64. Greater availability of metal and changing product mix of the Soviet economy suggest that a slightly higher figure for future years is plausible.

faith in the development of hydroelectricity. The fact that much of the expansion occurs in remote parts of Siberia where all of these considerations combine may suggest that it does not necessarily represent a deliberate attempt to penetrate Western markets.

In planning their methods of earning foreign exchange the Russians cannot ignore efficiency considerations. Where aluminum production falls on the hierarchy of possibilities open to them is unknown. However, by Western standards, it is hard to conceive of a firm contemplating production for export under the conditions faced by the Siberian industry. Presumably their power is inexpensive, but in every other respect the situation is unfavorable. They are forced to employ bauxite and nonbauxitic materials moved by land. Possible export markets can be reached only by expensive land transport over vast distances.

Aside from the question of costs, on which we remain ignorant, there is a further question of the extent to which market channels will be open to Russian metal. This is a matter of both private and public willingness to accept the metal. The trend toward integration to the fabricating stage is likely to progress further in Europe in the years ahead, thereby narrowing the proportion of the market in which the Russians could hope to sell ingot. As they are not in position to acquire their own fabricators, they could counter this trend only by exporting semis. They have in fact done so on a small scale already, but the marketing problems at this level become complex, normally requiring closer contact with customers, for dependability of supply, availability of technical services, and quality of products all become important in the semis market. Moreover, costs of shipping fabricated products are well above those for ingot, thereby compounding the Russian problem of distance. In less-developed countries, and especially those where government intervenes in trade relationships, the Russians may find an outlet for marginal amounts of metal. Those under foreign-exchange pressure might welcome the bilateral arrangements which can be made with the Russians.

So far the governments of Western European countries have been willing to accommodate a certain amount of Soviet metal. There is grave doubt that they would be willing to see this expanded in a substantial way. It can be expected that even if Europe remains a deficit area it will not wish to become dependent on Russian supplies of this basic industrial material. This re-

143

luctance is compounded by ignorance of long-term Russian plans and by the difficulty of bringing economical capacity into production on short notice if Soviet supplies should be cut off, as well as by sensitivity to the aspirations of Western firms to supply the demand. It must be remembered that there is some local production in nearly all Western countries. Government pressure in England already limits imports of Soviet metal. Belgian fabricators who have undersold EEC competitors, thanks to cheap Russian ingot, also have been subjected to heavy pressure.

The difficulties at both a commercial and government level surely are known to the Russians, and it seems unlikely that they will attempt to dump large supplies in Europe. More likely they will continue to press for some expansion, both there and in Japanese markets, but in such a manner as to avoid compelling stringent defensive measures. However, if the Soviets have in fact made a decision to use aluminum as a major permanent means of earning foreign exchange, then this decision should become apparent in the rate at which they push development of new capacity. The hydro potential of Siberia remains vast, and if still more aluminum capacity is added to that already announced before Russian consumption has a chance to grow up to it, then the implications of this for the West should be quickly drawn.

CHAPTER SEVEN

ELEMENTS OF THE COST STRUCTURE OF THE INDUSTRY

The preceding discussion in Chapter 4 of the technology of the aluminum industry provides the major ingredients for analysis of the various elements of the industry's costs. Because of its importance, electricity is reserved for subsequent treatment in Chapter 9. Insofar as the attention is on location, existing or prospective trade barriers must be considered an element of cost for those who wish to sell beyond them. The required rate of return on investment may be significantly higher in some circumstances than others and this also must be taken into account in considering location. The effect of these two considerations is dealt with in Chapter 10. The purpose here will be to indicate approximate magnitudes of other elements of cost and the influences which may cause them to vary.

No effort is made to estimate the actual cost of producing aluminum at various potential locations. That kind of effort would require such a long and tenuous chain of assumptions founded on so little information for so many hypothetical locations that it would provide no real answer. Rather an attempt will be made to indicate the kind of variations in cost which apply and their significance for the total. This will permit the assembly of at least a partial picture of locational advantage and of the effect of variation in different cost elements.

Bauxite is inexpensive to mine and transport. Because most of the value added in aluminum production arises at other stages of the process, bauxite so far has exerted relatively little influence on the location of smelting. Nonetheless, being a preferred raw material

145

for aluminum production, the distribution of bauxite reserves affects the industry's future logistics, and it may be possible for a country with a sizable and well-located deposit to gain some leverage over the location of subsequent stages of the industry. The existing pattern of bauxite production was described in Chapter 2. Reserves of ore present a very different pattern, however.

As with any mineral resource, world reserves of bauxite cannot be estimated with precision. This is perhaps especially true of bauxite, since its greatest occurrence is in tropical areas where exploration has been least thorough. The U.S. Geological Survey estimated that in 1963 world reserves amounted to 5,760 million long tons, nearly four times the 1950 estimate, and up by 2,500 million long tons over the 1958 figure. In addition they estimated potential lower-grade bauxite material at about 8,740 million long tons.[1] These reserves were widely distributed, with nearly 36 per cent in Australia, over 24 per cent in Africa, 12 per cent in the Caribbean, and 16 per cent in Europe. Lesser amounts were found in Asia and South America. On a country basis, Australia, Guinea, Jamaica, Hungary, Yugoslavia, Ghana, and Surinam were the leaders. Potential resources, as distinct from proved reserves, were concentrated even more strongly in Africa (especially Guinea), nearly 40 per cent, while China, Australia, Guyana, Cameroon, and Jamaica also were believed well endowed (Table 16).

As was noted, however, estimates of reserves have been adjusted upward to reflect new discoveries. Thus, the Australian reserves were listed at 600 million long tons in 1958 but in 1963 a single deposit at Weipa was reported as having over 2,000 million long tons, and subsequent reports have suggested vast Australian reserves of upwards of 3,000 million long tons or approximately equal to the estimated world total published in 1958. As these deposits are proved up, they will make Australia by far the most important potential future source of bauxite.

The two salient features of this distribution of bauxite reserves are: (1) although they are found largely outside the industrial centers, they are in areas to which non-Communist industrial countries should continue to have access, and (2) they are more than ample for the period of concern here. The availability of so many alternative sources of bauxite favors the bargaining position

[1] U.S. Geological Survey, *Estimates of World Bauxite Reserves and Potential Resources*, Professional Paper 475-B, by Sam H. Patterson (Washington: U.S. Government Printing Office, 1963).

Table 16. Estimated Reserves of Bauxite and Lower Quality Ores,
by Country

(million metric tons)

Area	Reserves of Bauxite	Marginal and Submarginal Resources
North America:		
United States	50	300
Total	50	300
Central America:		
Costa Rica	—	50
Panama	—	25
Total	—	80
Caribbean Islands:		
Dominican Republic	60	40
Haiti	25	—
Jamaica	600	400
Total	690	440
South America:		
Brazil	40	200
Guyana	150	1,000
French Guiana	—	70
Surinam	250	150
Venezuela	—	103
Total	440	1,520
Europe:		
France	70	190
Greece	84	100
Hungary	300	—
Italy	24	—
Rumania	20	—
U.S.S.R.	100	—
Yugoslavia	290	—
Other	9	30
Total	900	320
Africa:		
Cameroon	—	985
Congo	—	L(?)[a]
Ghana	254	—
Guinea	1,100	2,400
Malagasy Republic	—	25
Mali and Upper Volta	—	L
Malawi	—	60
Other	—	34.4
Total	1,350	3,500

Table 16 (Continued)

Area	Reserves of Bauxite	Marginal and Submarginal Resources
Asia:		
China (Communist)	150	1,000
India	58.1	200
Indonesia	25	10+
Iran	7	16
Malay	10	40
Philippines	—	28
Sarawak	5.6	—
Turkey	9.3	65
Other	—	10
Total	270	1,370
Oceania:		
Australia	2,060.3	1,190
New Zealand	—	20
Other	3	1
Total	2,060	1,210
Total for World	5,760	8,740

ᵃ L = Large.
Source: U.S. Geological Survey, *Estimates of World Bauxite Reserves and Potential Resources*, Paper 475-B, by Sam H. Patterson (Washington: U.S. Government Printing Office, 1963).

of industrialized countries in attempting to secure their supply. The existence of the Australian deposits provides a particularly stabilizing influence in this regard. The proved deposits of Australia, which are only now beginning to be exploited, would be more than sufficient to supply the bauxite needs of all non-Communist countries well beyond 1980.[2]

The aspiration of nonindustrial bauxite-rich countries to develop their resource and carry it through further stages of processing is favored by the fact that 4–6 tons of bauxite are required to produce a ton of metal and savings in shipping costs can be achieved if the material is processed nearer to the site of production. A favorable

[2] At a growth rate of 7 per cent per year from the 1960 figure of 20 million long tons, demand 1960-80 would total less than 900 million tons or less than one-sixth of known reserves. With known reserves on this scale and with more likely to be discovered, it is unnecessary to speculate on the quality of the potential reserves.

juxtaposition of bauxite and cheap power, along with political conditions conducive to investment, could provide a basis for the production of aluminum outside the present centers, given existing tariff levels. It must be remembered that while the distribution of bauxite reserves is of some relevance for the future location of the industry, the present pattern of production of bauxite is quite different from the distribution of reserves. Such current bauxite producers as Jamaica, Surinam, and France offer very limited opportunities for smelting operations based on existing power costs or potentials, but Guinea and Guyana may have greater possibilities.

Turning to the cost of bauxite, open pit mining methods for other ores commonly permit production at under $2 per ton under favorable conditions, exclusive of transportation, and should fall in the same range for bauxite. Washing and drying are inexpensive operations. Reimers figures the cost of mining, beneficiating (if any), and drying bauxite runs from $1.75–$4.50 per ton.[3] Royalties on bauxite in Guyana and Jamaica are 25 cents and 2 shillings respectively, with a sliding scale in effect tied to the aluminum price in the case of Jamaica.[4] Assuming they are similar in other exporting countries, royalties do not add greatly to the cost of metal.

Bauxite is mined and consumed almost exclusively by the aluminum companies. In this situation it is difficult to know how accurately price reflects cost. In the United States bauxite was valued at $7.28–$13.82 per long ton on a dried basis in 1963 at port of shipment while average value at the alumina plant was estimated at $17.15 per long ton.[5] Elsewhere a typical f.o.b. value for bauxite is estimated at about $7 per long ton.[6] The latter figure is in line with the figure given by Reimers of $6.50–$7 per long ton f.o.b. for Guyana bauxite of high quality while French bauxite f.o.b. mine was priced below $4 per long ton.[7] If we assume that 4–6 long tons of bauxite are required per long ton of metal, then the total cost of

[3] United Nations Secretariat, *Pre-investment Data on the Aluminum Industry*, ST/ECLA/Conf.11/L.24, January 28, 1963, by Jan H. Reimers, p. 17.
[4] United Kingdom Overseas Geological Surveys, *Royalties on Minerals in Certain Commonwealth Countries* (London: H.M.S.O., December 1961), p. 82.
[5] U.S. Bureau of Mines, *Minerals Yearbook 1963* (Washington: U.S. Government Printing Office, 1964), p. 285.
[6] Commonwealth Economic Committee, *Non-Ferrous Metals* (London: H.M.S.O., 1963), p. 42.
[7] Reimers, ST/ECLA/Conf.11/L.24, *op. cit.*, p. 17.

the required bauxite at the mine is commonly under $50 (often far lower) or less than 10 per cent of the sales price of the metal.

Transportation, of course, raises the cost of this raw material, but this can be minimized by location of alumina plants nearer to the ore. It is significant that the major metal-using centers in North America, Europe, and Japan are served by regional supply areas which have ample reserves. Japan is able to draw on the vast Australian deposits while Europe has ready access both to local supplies and to African sources. North America draws on the Caribbean area and the north coast of South America where total reserves of 1,100 million tons appear sufficient to last for several decades, even at presumed 1980 rates of utilization. Of the major consuming centers, North America is most exposed to eventual exhaustion of present sources of supply. However, West African bauxite (or alumina) can be tapped at little greater freight costs and Australian supplies for West Coast and Gulf smelters also are within reach.

With the major aluminum-producing and consuming countries increasingly dependent on imported ore, their advantages on bauxite costs vis-à-vis each other or vis-à-vis smelters in less-developed countries hinge more on transportation costs than on costs of producing the ore. The effect of transport costs is discussed below (p. 155 ff.). Suffice it to say at this point that the presence of bauxite deposits exercises comparatively little influence on the location of smelters. Such role as it does play is apt to be less a matter of economics than strategy on the part of major international firms who may seek to ensure goodwill in countries where they hold major mining concessions by locating some processing installations there. The leverage is likely to be most effective not on the location of smelters but rather on alumina plants where the companies' investment exposure is less and real economies are to be gained on subsequent transport costs.

ALUMINA

Alumina is produced almost exclusively by the aluminum firms. While some is used for other purposes and is offered for sale to such users, very little of that which is used within the industry is sold on an open market. Where firms are not self-sufficient in alumina, they may exchange metal for the raw material, as the Norwegian state firm commonly does. Where outright purchases are made, the price seldom is made public. Transportation may be important to the delivered cost of alumina, but as the contained metal content of alumina is over twice that of bauxite and its ton-mile shipping costs are no greater, it can move greater distances than bauxite.

Because alumina is largely an internal product of the major firms it is impossible to know precisely what it costs to produce it under various circumstances. One source has estimated cost at U.S. Gulf Coast plants at $55 per metric ton.[8] Prices cited by the U.S. Bureau of Mines are higher at $71.88 per metric ton for 1963 in the United States while imported alumina was figured at $59.03 per metric ton f.o.b. foreign port.[9] European prices cited by Bachmann ranged from $70–$76 per metric ton while American prices were thought to be lower.[10]

The principal elements of cost are capital charges and bauxite. At the alumina plant bauxite cost varies above all with transportation. Since 2–3 tons of bauxite are required per ton of alumina, significant transport savings can be had by locating near the source of the ore. The possible disadvantages of doing so are increased investment exposure and somewhat higher costs of other materials used in the process.

Investment in a large alumina facility is estimated by Reimers at $110–$150 per ton for a trihydrate plant, ranging $30 more for monohydrate.[11] Thus, a plant of 400,000-ton capacity capable of supplying a large 200,000-ton smelter would entail an investment of $50 million or so, apart from any harbor or townsite facilities required.[12] The significance of this for investment exposure is ob-

[8] U.S. Senate Committee on Public Works, *The Market for Rampart Power, Yukon River, Alaska,* 87th Cong., 2nd sess. (Washington: U.S. Government Printing Office, 1962), p. 155. It is unclear whether this figure allows for profit.

[9] *Minerals Yearbook, 1963, op. cit.,* p. 285.

[10] United Nations Conference on Trade and Development, *Aluminum as an Export Industry,* E/Conf.46/P/10, February 4, 1964, by Hans Bachmann, p. 15.

[11] See Chapter 4, p. 88. Elsewhere capital costs for a 350,000-ton plant designed to process Jamaican bauxite have been estimated at $143 per metric ton. See U.S. Bureau of Mines, *A Cost Estimate of the Bayer Process for Producing Alumina,* Report of Investigations 6730, by Frank A. Peters, Paul W. Johnson, and Ralph C. Kirby (Pittsburgh, 1966).

[12] The figures may be conservative. The Gladstone project in Australia is now quoted at $117 million for 670,000-ton capacity or about $175 per ton. However, this includes some mining and harbor development, and the plant is

151

vious, but the investment required is only about one-third of that needed to lay down a smelter consuming that amount of alumina. While alumina production is capital intensive, a company with the option of locating its alumina plant near the ore site and its smelter nearer to its market can enjoy the economies of shipping alumina rather than bauxite while holding its investment exposure to little more than one-fourth of what it would be if they located the entire complex near the ore site.[13]

At 10 per cent gross expected rate of return and 20-year life of plant, capital costs would amount to nearly $12–$17 per ton of alumina. A plant laid down in a less-developed country would probably cost more and investors would expect a higher return. Under the same depreciation assumption but assuming 20 per cent higher cost and a 15 per cent gross expected return, the capital charge would be about $7–$10 per ton higher—a sum which, if applied to transportation, would move the equivalent amount of bauxite over almost any probable desired distance.[14]

Aside from the question of transport costs on bauxite and capital costs, the other expenses of producing alumina do not strongly affect location. Detailed estimates prepared by the U. S. Bureau of Mines for a plant to process Jamaica bauxite show fuel, steam, and utilities at about $6 per metric ton while caustic ingredients cost under $3.[15] Caustic may be more expensive in some locations. With oil available at world prices, fuel need not be significantly higher elsewhere. Maintenance costs figures at over $3 in the United States would doubtless be higher at remote sites. Labor is expensive in the United States but very little is needed. Europe and Japan could save on this cost, but higher manning scales can be assumed for less-developed countries. Rough addition of these elements suggest that quoted price figures for alumina are reflective of cost.[16]

---

far from major industrial centers. The Fria project in Guinea is referred to as a $150 million investment for 480,000-ton capacity. In this case mining and transport costs are included. *Journal of Commerce*, October 3, 1963 (Fria), and *Wall Street Journal*, April 23, 1964 (Gladstone).

[13] If we must take power into account, the discrepancy is still greater.

[14] See Table 17. Note that the alumina normally must be transported from such a plant in less-developed countries to consuming centers in developed countries, so the penalty for moving bauxite instead of alumina is only approximately 1 additional ton.

[15] U.S. Bureau of Mines, Report of Investigations 6730, *op. cit.*

[16] *Ibid.*, p. 11. Total operating cost of producing alumina under the assumptions of $8 per ton of bauxite and labor at $2.30 per hour came to $42.81 per metric ton exclusive of any fixed charges.

All in all, there seems no compelling case for the location of alumina plants either at ore site or smelter. Given a secure investment climate in a less-developed country, the savings on transport and labor costs probably outweigh somewhat higher construction and maintenance costs and a probable higher price of caustic. However, the advantage is not overwhelming and the end result is affected also by political leverage. This is illustrated by recent and prospective projects. Thus, a new alumina plant is considered for Holland using imported ore at the same time that expansion continues near the ore sites. Reynolds expands at existing sites in the United States. In Australia a combination of very long hauls to prospective markets, and government pressure along with a secure investment climate, continue to favor local refining. Jamaica offers a good investment climate and convenient shipping point to numerous markets for alumina. Alcoa favors its good supplier, Surinam, with an alumina plant and Alcan does the same, perhaps with more trepidation, in Guyana. Harvey, faced with a long haul to its Pacific Northwest plants, chooses to refine its ore in the security of the Virgin Islands. Meanwhile the effort of the government of Guinea to expedite development of its resources was resisted by a cautious, bauxite-rich concessionaire—Alcan—and expansion-minded Harvey, backed by U.S. guarantees and aiming at prospective operations in Europe, appears to be willing to gamble on an ultimate refining operation there. On balance, the trend appears toward refining of bauxite into alumina nearer to the ore.

PRIMARY SMELTING

In addition to necessary plant with attendant tax, insurance, and maintenance costs, and the alumina discussed above, the principal costs of smelting include electric power, carbon electrodes, labor, and some electrolyte. While all of these costs can vary with location, the most significant variations are in capital costs, electricity, and labor.

Regional variations in power costs are treated in Chapter 9. Suffice it to say here that while power costs are an important component of the total and are the cost most subject to variation, several influences are at work which are likely to narrow the range of difference. The resulting spread between most favorable sites from the standpoint of power costs and the less favorable is likely to fall within a range 2–3 mills per kwh by 1980—a figure estimated to represent about $30–$45 per ton in cost of metal. Since sites with favorable power cost frequently have other disadvantages because of trading barriers, high capital costs, or transport costs which offset the advantage gained on power, the locational influence of power well may diminish.

The subject of trade barriers, insofar as they affect the location of smelters, is treated in Chapters 6 and 10. The effect of barriers at present levels and with likely trends in those levels is to protect small domestic production in less-developed countries while adding a penalty of $27–$49 or more per ton for would-be exporters to U.S. and EEC markets equal to 2–3 mills in power cost. Those non-industrial countries which have associate status with the EEC may avoid the penalty in that market, of course.

Let us look, then, at the remaining elements of cost. Materials necessary to the production of carbon anodes are readily available from petroleum coke and pitch. Reimers suggests prebaked anodes cost around $55–$70 per metric ton of carbon while the Senate Rampart study offers a figure equal to $76 per ton for anodes and pot linings.[17] With best practice consumption already under one-half ton of carbon per ton of metal produced, and likely to drop still further, a carbon cost on the order of $25–$30 per ton of metal can be expected in the future.

The cost of electrolyte is set at $16–$17 per metric ton of metal by the Rampart study and at $25 by Bachmann.[18] The figure is not likely to vary significantly from one area to another since the material is freely traded and can be synthetically produced. If the composition of electrolytes is changed by the introduction of lithium or other additives, it is likely that industrialized countries would have readier access to these exotic chemicals at some saving in cost.

Labor costs vary depending on local wage rates and manning scales. However, the problem is far more complicated if some mix of foreign supervision is required or if housing and other facilities must be provided. Under U.S. conditions the Rampart study estimated about 14 man-hours of labor and supervision per ton of

[17] Reimers, ST/ECLA/Conf.11/L.24, *op. cit.*, p. 31; and U.S. Senate Committee on Public Works, *op. cit.*, p. 155.

[18] U.S. Senate Committee on Public Works, *op. cit.*, Bachmann, E/Conf.46/P/10, *op. cit.*, p. 87.

metal produced.[19] Reimers allows for 18 man-hours at prebaked smelters.[20] In the most modern plants still fewer workers can be anticipated as automation proceeds.

Comparison of nominal wage rates can give a deceptive impression of unit labor costs. Manning scales in the United States commonly are lower than in Europe and Japan where labor is cheaper.[21] Moreover, fringe benefits provided are generally higher in proportion to wages outside the United States. In less-developed countries the costs of training labor must be added, and industrial labor employed in highly technical industries in such countries is able to command wages well above those prevailing elsewhere in their economy, especially if the employer is a foreign-owned firm. Nonetheless, unit labor costs are higher in the United States than in Europe and elsewhere because wages are higher and differences in manning scales do not fully compensate for this. In the United States the figure would be above $4 per hour and about 15 man-hours or $60 per ton of metal.[22] Japanese labor costs of approximately $40 per ton have been cited [23] while Bachmann suggests $30 per ton for an underdeveloped country.[24] For the most modern plants in developed countries with still lower manning scales, labor requirements are so low that it is unlikely that unit labor costs are significantly higher than in less-developed countries. In any case, the labor-cost saving likely to be achieved in a less-developed country is not great and may diminish as its wage scales rise and as labor-saving practices are adopted in developed countries.

TRANSPORTATION OF MATERIALS AND PRODUCTS

The movement of materials and products is an important element of cost in providing aluminum to end users. It has been noted that

[19] U.S. Senate Committee on Public Works, *op. cit.*, p. 155.

[20] Reimers, ST/ECLA/Conf.11/L.24, *op. cit.*, p. 31.

[21] *Ibid.*, p. 32, suggests below 20 in the United States, 25-30 in Europe and Japan, and 30-50 in underdeveloped countries.

[22] In 1962 the industry average was about $65 per metric ton for compensation of all employees. U.S. Bureau of the Census, *Annual Survey of Manufactures: 1962* (Washington: U.S. Government Printing Office, 1964), p. 40.

[23] Derived from *Oriental Economist*, May 1964, p. 320.

[24] *Op. cit.*, p. 87.

the principal ore, bauxite, tends to occur in tropical areas whereas consumption of metal is concentrated in temperate-zone industrial countries. Moreover, because of the cost significance of inexpensive electricity, materials may make a detour to remote power sources before being shipped as metal to consuming centers.

Transportation costs vary not only with the material shipped but with the mode of transport, specific conditions of the run, regulatory climate of the transportation industry, competitive situations, and other factors, making it difficult to generalize about them. Nonetheless, some of the dimensions of the problem can be suggested.

Based on the preceding discussion, the volume of materials required can be recapitulated. The amount of bauxite which must be shipped to produce a ton of metal varies with the quality and moisture content of the ore, but as a rough order of magnitude, 4–6 tons of bauxite are required, assuming it has been dried before shipment. Alumina amounts to about 2 tons of material per ton of metal. Finally, carbon, fluoride, etc., will amount to a bit more than ½ ton per ton of metal produced. Fabrication of metal produces scrap and reduces the weight of finished products below the amount of ingot input. However, if remelt facilities exist at the point of fabrication (and they commonly do), this loss has no significance for transportation costs because nearly all metal moved is ultimately used at destination.

At first glance, the minimization of transport costs would appear to favor processing at bauxite sources since 4–6 times the weight and still more in bulk must be moved if bauxite rather than metal is to travel. However, such a comparison is too crude for it ignores differences in cost of handling various types of cargo as well as the reverse transportation of other materials which may be entailed by processing bauxite at the mine.

A first processing to the extent of drying the bauxite at the mine is common practice. If fuel is available for raising process steam and for calcination, there are then transport advantages to refining bauxite into alumina at or near the mine since this reduces weight by half or more while only small weights of soda and other materials are consumed. This has the additional advantage that alumina is shipped as a bulk material and is easily handled so as to be no more expensive to move on a per ton basis than bauxite.

There is no transport advantage to be had from moving fuel to the bauxite source for smelting ore to metal for, in addition to the carbon and fluoride used in reduction and the soda and fuel used in the alumina stage, about 4–5 tons of coal would be required to produce enough power to smelt a ton of metal and the metal would still have to be shipped to market.[25] Thus the advantage lies with moving the bauxite or, far better, the alumina.

Cargoes that move as bulk materials can be handled more cheaply per ton than more refined products. This is particularly the case where specially designed carriers and equipment are available to handle the materials concerned. Thus, bauxite and alumina move readily on a bulk basis while metal, and fabricated products to an even greater extent, require different handling and are more expensive to ship. In part this reflects the greater diversity in destination of metal as compared with materials while, in the case of fabricated products, it also reflects the greater care required in handling to avoid damage.

Apart from the weight and bulk of material handled and the type of care that it requires, certain established principles of transport economics should be borne in mind as general guides. Generally speaking, ton-mile costs decline with distance because the relatively fixed costs of time spent in terminals and the loading and unloading charges are spread over a greater number of miles. Also costs generally decline on a ton-mile basis if larger carriers are employed, because both capital and labor costs per unit of capacity are lower. Finally, the hierarchy of costs by various means of transport favors ocean transport where feasible, followed by barge, train, and truck in order, under most circumstances. The difference in ton-mile costs by these various means under the circumstances for which they are best suited are of discrete orders of magnitude, although in some cases the vagaries of pricing may obscure this fact.[26]

Aside from these generalizations, the specific circumstances of the run greatly affect the economics of transportation problems in various ways. A most obvious example is the ice problem en-

[25] Estimate of fuel requirement was based on 15,000 kwh and 28,000,000 btu per ton of coal and 8,000 btu per kwh.

[26] For long-distance hauls and using American figures for inland transport, the rough figures per ton-mile for comparable bulk commodities would be on this order: steamship .2 cents, barge .4 cents, train 1.25 cents, truck 6.5 cents (National Association of River and Harbor Contractors, *Waterways of the United States* [Washington, 1961], p. 33).

157

countered by Alcan on the St. Lawrence which forces it to stockpile materials during the ice-free season. This same circumstance leads it to seek alternative uses for its shipping during the off-season and has contributed to its role as the largest shipping line in Canada.

This one illustration only hints at the diversity of special circumstances. Some harbors are too shallow for large carriers and lighters must be used. Port equipment and charges vary enormously throughout the world. In some cases the port facilities are owned and maintained at company expense and in others the public pays for them. Availability of return cargoes can make a great difference in cost. Often rates are quite different in one direction than for a run in the opposite direction. Rail rates commonly are set by government-owned companies or by government regulatory agencies. They may favor certain classes of goods or hauls in one direction or may make concessions to ensure competitiveness of a domestic industry. They also may set rates which are influenced by competitive means of transportation and, in consequence, have scant relation to cost.

For all of these reasons it is dangerous to refer in any general way to simple ton-mile costs of moving materials and products. Often the companies own their carriers, employ them in diverse ways, and arrive at cost figures based upon quite different accounting assumptions. The best that can be done to get at least an order of magnitude of transport costs is to cite some typical and important hauls and the costs estimated for them. Some examples supplied by industry sources are shown in Table 17.

Table 17. Transport Costs for Materials and Products in the Aluminum Industry for Selected Runs

Mode of Transport	Run	Distance (miles)	Dollars per Metric Ton	Cents per Metric Ton-Mile
		Bauxite		
Steamship	Surinam–Holland	4,600	6.50	.141
"	Guyana–Quebec	3,157	9.00[a]	.285
"	Surinam–Texas	2,810	7.00	.249
"	Malaya–Japan	2,802	5.00	.178
"	Jamaica–Texas	1,290	1.75	.136

158

Table 17 (Continued)

Mode of Transport	Run	Distance (miles)	Dollars per Metric Ton	Cents per Metric Ton-Mile
	Alumina			
Steamship	Guyana–British Columbia	6,390	13.00	.203
"	Texas–Oregon	5,620	9.75	.173
"	Jamaica–British Columbia	4,937	9.50	.192
"	Jamaica–Norway	4,420	7.80	.176
"	Japan–Oregon	4,390	6.40	.146
"	Guinea–Norway	3,595	7.40	.206
"	Jamaica–Quebec	2,530	9.25[a]	.366
"	Jamaica–Louisiana	1,210	2.25	.186
Rail	Alabama–Oregon	3,144	13.51	.430
"	Alabama–New York	1,724	8.75	.507
"	Alabama–Tennessee	535	5.90	1.102
"	Arkansas–Arkansas	50	1.05	2.094
	Ingot			
Steamship	British Columbia–Australia	7,000	25.00	.357
"	Quebec–Argentina	6,300	30.00	.476
"	British Columbia–Japan	3,800	25.00	.658
"	Norway–New York	3,600	7.15	.198
"	Quebec–Holland	3,135	18.50	.590
"	Quebec–United Kingdom	2,815	17.80	.632
"	Norway–United Kingdom	1,100	10.00	.909
Barge	Texas–Iowa	1,900	8.95	.470
"	Louisiana–Ohio	797	8.00	1.004
Rail	Washington–Illinois	2,321	23.65	1.019
"	Texas–Tennessee	1,183	23.15	1.957
"	Washington–Washington	20	2.10	10.470
Truck	Indiana–Indiana	120	5.75	4.770
	Coiled Sheet			
Steamship	Ontario–Argentina	6,400	66.00	1.031
"	Washington–Japan	4,390	38.15	.869
"	Ontario–Holland	3,300	42.00	1.273
Rail	Iowa–California	2,111	41.45	1.963
"	Tennessee–New York	728	18.50	2.544
"	Tennessee–Georgia	200	6.70	3.362
"	Iowa–Illinois	154	6.30	4.080

[a] Including topping, unloading, and 20-mile rail haul.
Source: Data provided by several aluminum companies.

The foregoing information illustrates the difficulty in generalizing about transport costs. However, it is noteworthy that the steamship rates for bauxite and alumina do not appear to differ significantly for comparable runs and the rates generally fall in the range of 1½–2 mills per ton-mile. Steamship rates for ingot tend to be 2–3 times the rate for raw materials. The rates quoted for ingot are liner terms and it is possible that they would be somewhat lower if regular runs of large tonnage could be made.[27]

[27] This possibility is of interest in considering an export smelter in a less-developed country regularly exporting to, say, a major fabricator in a developed country. In this case some economies of shipload lots at charter rates or equivalent might be feasible and this could hold down cost. However, ingot would incur significantly greater handling charges and would never justify the largest scale carriers. Moreover, regular return cargo might be hard to come by. Thus, costs would remain higher than for materials.

This conclusion hardly should be considered surprising since it is not unreasonable to suppose that existing rates reflect real cost ratios. However, an illustration may serve to reinforce the point. If all of the output of a large smelter (200,000 tons) were shipped point-to-point over a run of 5,000 miles, it would require only a single ship in the 15,000-ton class to handle the traffic. According to information supplied by shipping firms the running costs per ton-mile for such a ship would be over one and one-half times the costs for a large bulk carrier of about four times this size. In addition, the cargo ship would spend perhaps three times as many days in port. Finally, even with the most advanced concepts of cargo handling, ingot cannot be loaded and unloaded as efficiently as alumina, which can be poured in and sucked out of the hold by pneumatic equipment; per ton handling charges will be far higher for the metal. Port costs on general cargo, while highly variable, often amount to well over one-half of total shipping costs and even with standardization on best practice cannot be brought into the same range per ton as bulk cargo. If the ingot is shipped to diverse destinations, the economies of shipload lots could be had only at the cost of flexibility.

While future cost reductions might be expected for a hypothetical point-to-point ingot run, the future economies in handling bulk cargo seem likely to prove more impressive, thereby frustrating any shift in favor of ingot movement. It is tempting in this connection to cite developments in oil tankers where mammoth-sized ships (100,000-300,000 tons) are now appearing, bringing dramatic reductions in both capital and operating costs. Although this is sure to have some carry-over into the bauxite and alumina trade, these possibilities cannot be translated directly. The volume of materials employed in aluminum reduction is on a much smaller scale. A large 600,000-ton alumina plant sited near an ore source could ship its entire output in a single 50,000-ton carrier assuming 12 trips per year. Traffic on this scale would not justify expensive harbor facilities to accommodate the monster carriers of the size contemplated for the oil trade. Indeed, vessels of the largest size will rarely be able to put into port but rather will lie in deeper water to pump cargoes on and off—an option not open to dry-bulk carriers which must tie up at the dock or employ lighters (an expensive procedure). However, where harbors and volume of traffic permit, the effect of larger carriers will be to facilitate the movement of bauxite as well as alumina, in both cases to the disadvantage of

Nonetheless, if ocean transport is feasible, the above rates suggest that it is no more expensive to move 2 tons of alumina than its contained aluminum in the form of a ton of metal.[28] Since bauxite and alumina move at about the same cost per ton-mile, there is a clear transport advantage in refining the bauxite into alumina before long-distance shipment.

Few examples of barge rates are included above and costs can be expected to vary greatly on different waterways. In the U.S. Mississippi–Ohio system published rates are not the best guide because owned or chartered barges on regular runs can do better. A figure on the order of 4 mills per ton-mile can be taken as representative for bulk commodities.

Rail rates represent a significant jump. For long hauls in the United States rates of over 10 mills are to be expected. The favorable rate shown for transportation of alumina to the Northwest is available in part because of the direction of the traffic. While rates in Europe were not obtained, it is understood that they tend to be higher because runs are short and equipment is of smaller capacity. Special rates are understood to aid the competitiveness of some of the inland plants in Europe which must move bulk materials overland.

While there is little to choose from between moving alumina and ingot, there is again a clear choice between ingot and fabricated products. For the examples chosen and for comparable distances and modes of transport it appears about twice as expensive to move a ton of coiled sheet as a ton of ingot. Coupled with the higher tariff on fabricated products, this does much to explain the fact that trade heretofore has concentrated more on ingot than fabricated products.

From these examples several conclusions can be drawn. First is the marked advantage of ocean transport for moving bulk materials in particular. Second is the fact that cost is not proportional to distance, especially for more refined products and for land hauls.

the aspirations of less-developed countries to engage in further processing. Any significant change in the relationship to transport costs seems more likely to favor bulk carrying costs than finished products, thereby reinforcing the position of developed countries as locations for reduction plants.

[28] The glaring exception to this generalization in Table 17 is the Norway-New York traffic. According to a shipping industry informant, this low rate is not profitable and is due solely to the lack of cargo in this direction. Note that traffic in the opposite direction moves at 2-3 times this rate.

Third is the fact that for comparable hauls and mode of transport there is a sharp difference in cost per unit of weight among raw materials, ingot, and fabricated products. These factors work out in such a way as to favor the long-distance transport of alumina to metal-consuming centers rather than bauxite or metal, and the transport of ingot in preference to fabricated products. On the other hand, it is generally cheaper, but not by a wide margin, to transport ingot rather than bauxite to metal-consuming centers.

The fact that metal does not move as cheaply as the amount of alumina required to produce it is of great significance, for it gives a surprising advantage to the location of smelters nearer to markets in developed countries and is a disadvantage to the prospects for smelting in less-developed countries.

Given the fact that materials must be moved to power sources and metal must be transported to markets, it is worthwhile to note the cost of transport in terms of equivalent differences in power cost. A one-mill difference in power cost can be expected to affect ingot cost by $15 per ton (assuming 15,000 kwh per ton). At 1.5–2 mills per ton-mile this amount would be the cost of moving 2 tons of alumina 3,700–5,000 miles or a ton of aluminum 3,000 miles at 5 mills per ton-mile, both rates appropriate for long ocean hauls.

To compensate for differences in transportation costs only, a smelter in a less-developed country using local ore would need slightly lower power costs than prevail in its export market if it is to ship metal rather than alumina to that market. A third country could import alumina for reduction more advantageously the closer it lies (en route) to the market.

CHAPTER EIGHT

TECHNOLOGICAL PROSPECTS

$\mathbf{J}$ust as the recent trend in the location of aluminum production has been influenced by the changing size and pattern of demand and by new sources of raw materials, so it also may be affected by potential technological changes in the industry. While it is impossible to discern the full effects of problematical future changes in technology where cost relationships remain unknown, some speculation concerning the prospects for new technology, the direction of its impact, and some illustrative suggestion of the magnitude of its effect under certain assumptions appears to be worthwhile.

The two chief forms which technical change may take are (1) improvements in the basic Bayer-Hall process and (2) the replacement of this process with a radically different reduction technology. In addition, there is the possibility of recovering aluminum commercially from nonbauxitic materials, either by refining them to alumina by methods other than the Bayer process or by employing them in a direct reduction process. The implications of these possibilities must be considered.

Existing technology can be expected to improve through ordinary engineering advances. The cost of mining and transporting raw materials will be moderated by improvements in earth-moving and in bulk-handling techniques. This will stem mostly from the use of larger units of equipment. It should be borne in mind that the mining cost of bauxite already is very low. Hence the potential savings in this area can have only a limited effect on the cost of metal. In addition, the tendency previously observed for alumina plants to locate nearer to ore sources will help to minimize the importance of improvements in transportation on cost.

The Bayer process of alumina production already is a highly efficient one. It may require modification in response to the changing nature of the ores employed, and, in addition, the possibilities of improvements in materials handling and heat exchanger design should not be ignored. The Fria plant in Guinea provides an interesting design departure by using leaching at atmospheric pressure, thereby simplifying design and maintenance of the plant, but at the cost of greater consumption of soda.[1] In general, however, the Bayer process seems very well developed and is unlikely to alter much in the period of concern.

More significant improvements may be expected at the reduction stage and perhaps in fabrication. While the latter may be very important in restraining the cost of fabricated aluminum to industrial users, it has limited significance for location of primary production.[2] So far as reduction is concerned, economies may be sought by improving electrical efficiency, utilization of work force, and efficiency of equipment. These possibilities are interrelated since, as was pointed out earlier, improved electrical efficiency may involve greater spending on capital and labor, while economies in labor can be achieved primarily through greater capital spending.

Some industry sources have considerable hopes for automation of the reduction plant in the belief that this will save on direct labor as well as other operating expenses. Automation will permit a closer approach to the maintenance of optimum conditions in the reduction cell, with attendant improvements in electrical efficiency and carbon consumption. Automatic control of feeding, carbon adjustment, and tapping is surely feasible and it is likely to occur

---

[1] United Nations Secretariat, ST/ECLA/Conf.11/L.24, January 28, 1963, *Pre-investment Data on the Aluminum Industry*, by Jan H. Reimers, p. 3.

[2] The most promising innovation in metal fabrication is the development of various techniques to bypass the ingot stage. Breaking down an ingot requires powerful equipment which normally implies a larger scale than is justified to process the output of a small smelter. Most attention has focussed on continuous casting machines which cast the molten metal in a form more easily worked by lighter equipment. So far, such techniques have not been widely adopted by the larger producers, but, as additional experience is gained, continuous casting may prove more popular, especially in circumstances where heavy, large-volume equipment is not justified. This would act to the advantage of small national industries in less-developed countries.

164

in major new plants before the end of the period of concern. It will permit plants of otherwise similar design to produce more metal (perhaps over 20 per cent more) with the same power and less labor but somewhat greater capital costs.[3] At the same time it should be recognized that the potential labor–cost savings through this means are limited. Modern French plants already achieve potroom labor inputs as low as 4½ man-hours per ton and Reimers figures 8 hours per ton for a modern plant of medium size.[4] There is some feeling that reduction below this level would entail small savings and would carry the risk of the potroom being under-manned in case of emergencies there. Because the costs of automation remain unknown, the prospective net saving in cost per kg of metal is difficult to estimate, but it could be significant.

Electrical efficiency may be further improved, even without automation, by means of improved electrolyte and cathode materials. With the electrical efficiency of cells at 30–35 per cent, most of the energy is being used to overcome resistances.[5] Voltage loss at the cathode can be reduced by .2V through use of titanium diboride embedded in the cathode, while the addition of lithium to the electrolyte permits the increase of current density and improves heat dissipation. In a test of lithium additives, it was shown that cell output could be raised by 8.2 per cent, power consumption could be decreased by 2.5 per cent per unit of output, and carbon consumption per unit of output by 2.5 per cent. Use of lithium additives along with the new cathode materials yielded a 37.8 per cent increase in production of the test pot and a 22 per cent reduction in carbon consumption per unit of output.[6]

It should be noted that improved electrical efficiency brought about by the use of electrolytes and cathodes of greater conductivity may be used either to produce the same amount of metal with less current or to produce more metal with the same equipment by means of greater electrical efficiency and higher current density. These possibilities permit a two-way savings in cost because of the use of less electricity per unit of output and less capital per unit of output.

Lithium additives and boride cathode materials currently are

[3] Based on discussion with technical people in the industry.
[4] Reimers, ST/ECLA/Conf.11/L.24, *op. cit.*, p. 29.
[5] *Chemical Week*, "Aluminum," September 15, 1962, p. 90.
[6] *Ibid.*, p. 90.

quite expensive, so the net savings are somewhat less impressive. Nonetheless, one major American producer, Reynolds, publicly announced its intention to expand output by 20 per cent at existing plants, chiefly by coaxing greater production through greater current densities in existing facilities rather than by the addition of new pot lines.[7] In brief, then, prospective improvements in the Bayer-Hall process will act in the direction of reducing power requirements per unit of output and of increasing the capacity of existing plants or of future plants incorporating these improvements; this is the equivalent of a reduction of both operating and capital costs per unit of output. In addition, the possibilities of smelter automation offer the opportunity for substitution of capital for labor inputs under appropriate conditions.

Because many of these improvements, especially those affecting electrolytes and cathodes, can be employed to increase capacity in existing plants at low cost, they offer a partial substitute for new plant capacity. Moreover, by reducing the amount of power per unit of output they diminish the importance of low-cost energy. Insofar as capital requirements per unit of capacity are reduced, the extent of exposure occasioned by investment in less-developed countries likewise is reduced. Automation, however, raises capital requirements to the disadvantage of such areas, and it can be assumed that this as well as other trends toward more sophisticated technical requirements will favor location in the major industrial countries.

Upgrading of capacity at existing smelters based on these possibilities already is well under way, permitting the postponement of decisions on new locations and enhancing the competitiveness of existing plants and of locations which otherwise would suffer from fatally high power costs. It can be expected that these improvements will be widely adopted throughout the industry over the

[7] Interview, R. S. Reynolds, *American Metal Market*, May 28, 1963. This may indicate that in their view the cost relationships favor such improvements to the Hall-Héroult process as the least expensive way of increasing output. This conclusion need not follow, however, because the same announcement foresaw that future new plants would utilize radically different technology. Hence, company strategy might have favored somewhat higher present costs of increasing output at existing plants pending perfection of the new technology instead of expanding output by building new plants already believed obsolescent. See also *Metal Bulletin*, July 5, 1963, p. 20. In a subsequent announcement, Reynolds has gone on record as planning a coal-fired Ohio Valley plant, presumably of conventional design.

next few years. Because knowledge of the cost of effecting the improvements is not available to persons outside the industry, it is impossible to specify more fully the net effects on total cost and on location.

Throughout the history of the aluminum industry there have been persistent attempts to find substitutes for bauxite as the basic source of metal. These attempts are prompted by the knowledge that aluminous materials are of common occurrence while bauxite supplies in developed countries are limited. To date, however, the only significant use of ores other than bauxite is found in the U.S.S.R. where the shortage of bauxite and a state policy of relative self-sufficiency have resulted in the use of nepheline.

In the West the chief focus has been on attempts to make alumina from clays and from coal mine tailings. One process has been pursued through the pilot plant stage by Olin Mathieson which claims good success with clay or coal mine shale. It involves the leaching of the material with sulphuric acid to bring aluminum and iron into solution, separation of aluminum sulphate by crystalization, and recovery of aluminum and sulphuric acid by decomposition of the crystals. The significant part of the process is a technique for securing large crystals which can be separated easily from the liquor. Clays and shales with only 20 per cent alumina content can be treated by this process.[8]

Another process is proposed by Anaconda and is the basis of announced plans for a future alumina plant. In this case clay is calcined and leached with hydrochloric acid to dissolve the aluminum and iron. The solution is concentrated and roasted, leaving ferric oxide and alumina, the hydrochloric gas being recycled. The iron can then be separated, much as in the Bayer process, by caustic soda digestion or by a soda sinter process.[9]

The U.S. Bureau of Mines has examined both sulphuric acid

[8] Finn B. Domaas, "New Methods, New Ore Sources Attract Aluminum Industry Research Efforts," *Engineering and Mining Journal*, October 1961, pp. 106-9. North American Coal Company (NACCO) and Vereinigte Aluminium-Werke (VAW) are associated with a similar process (*Chemical Week*, "Aluminum," September 15, 1962, pp. 92-93).

[9] Domaas, *op. cit.*, p. 109.

and hydrochloric acid processes for extracting alumina from clay. They conclude that the processes, while technically feasible, are not competitive with the Bayer process under current conditions.[10] The literature is full of descriptions of various substitutes for the Bayer process using both bauxite and nonbauxitic materials and both acid and basic processes.[11] It is significant, however, that none of these processes is used commercially and some tried under straitened circumstances (as in wartime) have been abandoned as soon as possible in favor of the Bayer process. During the period of concern, if the electrolytic smelting process remains the standard, then the alumina feed material is likely to continue to be made by the Bayer process from bauxite. Departures from this may occur, however, where an especially inaccessible but otherwise advantageous power site occurs, or for reasons of national or company self-sufficiency. Anaconda, which presently must rely upon purchased alumina, recently has announced plans to go ahead with an alumina plant based on clay.

Should the alumina stage be bypassed, it is conceivable that local nonbauxitic materials might be used as a source of metal. The obvious advantage would be their widespread availability in developed countries. It is likely that they would be available near power sites—certainly in instances where coal is used to generate the electricity for reduction—but also very likely in those remote areas where surplus hydropower also might be available. However, the prospective savings on transport costs in these cases must be balanced against the fact that nearly all nonbauxitic materials are lower in aluminum content and therefore the mass to be smelted per unit of metal recovered would be greater than in the case of bauxite.

[10] U.S. Bureau of Mines, *Methods for Producing Alumina from Clay: An Evaluation of the Sulfurous Acid-Caustic Purification Process*, Report of Investigation 5997, by Frank A. Peters, Paul W. Johnson, and Ralph C. Kirby (Pittsburgh, 1962), and U.S. Bureau of Mines, *Methods for Producing Alumina from Clay: An Evaluation of Five Hydrochloric Acid Processes*, Report of Investigation 6133, by Frank A. Peters, Paul W. Johnson, and Ralph C. Kirby (Pittsburgh, 1962).

[11] For example, see also United Nations ECAFE, *Bauxite Ore Resources and Aluminum Industry of Asia and the Far East* (Bangkok, 1962), pp. 266-67; United Kingdom Overseas Geological Surveys, *Bauxite, Alumina and Aluminium* (London: H.M.S.O., 1962), pp. 51-58, and T. G. Pearson, *The Chemical Background of the Aluminium Industry* (London: Royal Institute of Chemistry Lectures, Monographs and Reports, 1955), pp. 83-92.

NEW REDUCTION TECHNOLOGIES

Efforts to supplant the Hall-Héroult process have been the subject of much research in the industry over many years.[12] Except for special wartime circumstances, no commercial substitute for the process has yet been used. At present, however, several companies are experimenting with new direct reduction processes or other techniques, both in the laboratory and in pilot plant operations. It is still too early to know whether these attempts will succeed in besting the ever-improving Bayer-Hall process. Laboratory results are very closely guarded and press releases sometimes appear to be less a function of real progress than of the companies' felt public relations needs. Of the two companies known to have pilot plants in operation using quite new reduction techniques, Alcan has been very cautious in its claims and Péchiney still has not scheduled commercial application of its technique. Reynolds, not known to have a pilot plant in operation, once promised that its next new plant would feature an unspecified new process, but at the same time the date of construction was pushed back by simultaneous announcement of plans for considerable expansion via improvements in existing conventional plants, and a subsequent announcement of plans for a new plant omitted any mention of a new process.[13]

Two different approaches to reduction are associated with Alcan and Péchiney. Numerous other processes have been described to produce aluminum alloys.

The Alcan approach involves the reduction of bauxite (presumably other aluminous materials would work also) with carbon at high temperatures to a molten state which then comes in contact with aluminum trichloride ($AlCl_3$) in an unpressurized reactor at 1,000°–1,200° C. A volatile monohalide ($AlCl$) forms and flows to a condenser, leaving the impurities behind. In the condenser a pool of molten aluminum is maintained at a lower temperature of 700°–800° C. An impeller throws up a shower of this metal to cool the incoming monohalide, causing the initial reaction to reverse and to precipitate metal from the $AlCl$, while

[12] For early description, see Junius Edwards, Francis Frary, and Zay Jeffries, *The Aluminum Industry and Its Production* (New York: McGraw-Hill, 1937), Vol. I.
[13] See footnote 7 above.

the AlCl$_3$ is recycled. A key feature of this process is the use of a body of molten salt at 200°–400° C in a separate compartment of the condenser to absorb heat from the aluminum bath. The salt in turn has cooling coils to remove its heat. This two-stage procedure prevents contamination of the aluminum from any attempt to maintain cooling coils directly in the bath.[14]

No significant details on the operation of Alcan's pilot plant have been released. The savings were anticipated to be a 50 per cent reduction in capital cost (in part because alumina plants are eliminated) and lower operating costs as a consequence of lesser requirements for labor and for such materials as caustic, cryolite, and electrodes. Apparently power requirements are not expected to differ greatly from the Hall-Héroult process.[15] So far the reports on this process have been less than enthusiastic in the pilot operation, but it is still too early to know whether the earlier hopes will be realized. Critics have noted that the process requires melting the entire mass at high temperature and considerable waste of heat. Moreover, raw bauxite does not make the best charge and some advance processing (pelletizing) is advisable. Another problem is to find suitable and economic refractories to handle the very corrosive halides involved. If this problem can be mastered there is good hope for the process once it has been scaled-up for production.[16]

Péchiney has described two different approaches: a carbothermic process and a nitride process. Unlike the Alcan process the carbothermic process cannot be termed a direct reduction technique since it relies upon the prior refinement of bauxite into alumina. This need not be accomplished by the Bayer process, however, for Péchiney has elaborated a different technique for smelting bauxite to get alumina.[17] Using the carbothermic process, the Al$_2$O$_3$ and carbon are charged into an arc furnace at 2,500° C and react into a molten mass consisting of aluminum carbide (Al$_3$C$_4$) and some carbon and aluminum. The Al$_3$C$_4$ is removed,

[14] Domaas, *op. cit.*, pp. 105-6; and *Chemical Week*, "Aluminum," *op. cit.*, p. 92.

[15] Domaas, *op. cit.*, pp. 105-6.

[16] In discussion with a senior technical officer of Alcan it was suggested that most of the technical problems have been solved and those which remain relate to scaling-up for production. It was further stated that whereas hoped-for cost savings through improvements in existing techniques are to be measured in mills per lb, the savings from the new process would be in cents per lb. This is the most optimistic appraisal of the new technique that the author has encountered from a responsible source.

[17] Domaas, *op cit.*, p. 109.

heated, and decomposed and the aluminum is condensed and collected. The nitride process involves the thermal production of aluminum nitride (AlN) and the subsequent condensation of aluminum from this vapor.[18] Although Péchiney has experimented with these techniques at the pilot plant level since 1960, there remains no evidence that they are about to be adopted commercially, so apparently they have not yet been shown to be commercially feasible.

Should the Hall-Héroult process be replaced by any of the foregoing processes, the consequences for location might be considerable. Several features of the new process account for this, namely: (1) the proposed technology would not make use of Bayer-process plants, (2) it appears that the reduction itself could be accomplished economically at a smaller scale than with the electrolytic process, (3) investment per unit of capacity should be substantially less than with present technology, and (4) the new processes are likely to require more sophisticated technical and operating personnel (albeit even fewer employees) than present smelters.

Elimination of the Bayer plant will have several effects. It will be recalled that Bayer plants are expensive and must be built to large scale to be economical. At the same time they reduce by one-half or more the weight of the raw material which must be transported to the smelters. Elimination of the Bayer plant would mean that a small company or a country not wishing to be dependent upon others for its supply of alumina no longer would have to operate a Bayer plant of uneconomically small size with its attendant heavy capital outlays.

In addition it should be borne in mind that none of the major industrialized countries of the non-Communist bloc, except France, is in a position to meet its bauxite needs from domestic production; the others must import. Conceivably aluminous materials other than bauxite could be employed in the new processes, and there might be some attraction stemming from the possibilities of using domestic sources of such materials as clay or mine tailings. How-
low cost. Transportation of so bulky a material over long distances
ever, bauxite remains a desirable feed material and can be had at
would mean abandonment of the recent trend toward initial processing into alumina near the ore source—a trend which has reinforced the location of smelters in industrial countries.

[18] United Nations ECAFE, *op. cit.*, pp. 31-32.

171

A smaller economic size of plant and lower investment costs per unit of capacity are alike in their effects on location. They make it possible to go into business on a smaller scale and with less capital outlay. This may be of importance to individual firms denied entry by present scale and cost, and the same is true of nations which may aspire to enter production. From the standpoint of foreign investors these two features diminish the investment exposure required for any given amount of capacity in a less-developed country and permit readier dispersion of this risk at smaller plants in different countries. Moreover, the smaller economic scale makes feasible the exploitation of smaller power sites or local bauxite supplies and means that regional or national markets of limited size may still be served by a plant of economic scale.

The higher level of technical sophistication called for by the new processes clearly favors the existing industrial countries where suitable technical personnel are to be found. It is expensive to maintain skilled personnel from industrial countries in less-developed areas over long periods of time. Additionally, while existing technology is widely available to any who choose to enter the industry, the new techniques will be protected by patents and closely held expertise for some time and will be made available only upon terms which the major companies find satisfactory.

The consequences of these characteristics for different parties will vary. If the prospective technology is considered as a replacement of existing technology in industrial countries, it loses part of its allure. The advantage of smaller economic scale is not great for such countries because they have ample capital and their large markets can absorb the output of large plants. Smaller scale would permit a somewhat closer regional fit between production site and market and might permit the exploitation of resources in the interior of, or production to meet the needs of isolated portions of, such countries.[19]

The more serious drawback for the new processes in industrial countries is the increased transportation cost entailed in moving bauxite rather than alumina. This would diminish the attractive-

[19] There also are possible implications for the structure of the industry where smaller-scale and lower-entry costs might permit fabricators who aspire to have their own source of metal to build smelters. The importance of this under the existing market structure is not great because there are few fabricators of significant size in industrial countries who are not tied to primary producers. Moreover, under circumstances of easier entry, the primary producers would be expected to be more sensitive to the needs of fabricators.

ness of the new process in the Ohio Valley, for instance, and would reinforce the importance of transportation as a location factor. Nonetheless the new processes, if they come anywhere near expectations, should be able to bear the added transport costs and compete effectively with present technology within industrial countries.[20]

How would the new technology affect the incentive of the major international firms to invest in smelters nearer to the ore sources from which they could serve markets in industrial countries? In this instance they would save on transport costs over similar plants located in industrial countries. The fact that scale and investment per unit of capacity would be smaller would permit reduction of their investment exposure, and it also might be dispersed over several countries. Nonetheless such exposure would be great compared with locations within industrial countries and the firms would be reluctant to undertake it. The aforementioned savings on transport costs alone probably would not justify it. In few cases could the bauxite be processed at site, and, if it had to be transported, the additional cost of distance to the home-based smelters of industrial countries might prove no great burden. Moreover, the fact that the alternative would be to ship metal at much higher rates diminishes the pull of the bauxite.

The new processes hold considerable promise for nonindustrial countries that aspire to have their own national industry. The technical hurdle would have to be surmounted by some liaison with a major firm to ensure availability of licenses and know-how. The lesser scale and lower capital requirements would be a real boon in cases where their domestic markets, capital and exchange funds, and power or bauxite resources are limited. It would mean that such countries nonetheless might build a domestic industry without the penalty of high cost which they so often now incur.

The proposed technology would show to best advantage in cases where low-cost power and bauxite occur at the same site. In practice there are few such situations around the world and none of large potential. If power and bauxite do not coincide, then the

[20] As an illustration, if cost savings are measured in "cents per pounds," then this implies as a minimum something over $40 per ton. If $4\frac{1}{2}$ tons of bauxite are transported 5,000 miles at 2 mills per ton-mile, the cost ($45) is only $25 more than the cost of transporting 2 tons of alumina under the same conditions. Thus, added transport costs would not kill the new process. However, still more distant bauxite sources (Australia, for example) would be disadvantaged.

added transport costs which are entailed by moving bauxite instead of alumina must be more than offset by the operating and capital cost savings of the new process. If the savings prove to be of the order anticipated, however, it should be advantageous to accept the added transport costs even at sites distant from the ore sources.

On balance, the prospective technology appears to tilt the scales somewhat toward less-developed countries, both for local market production and for export. Their only new disadvantage is in technological know-how and this can be circumvented by association with a major firm. They gain on the transport equation and in the factors associated with investment, whether using their own or foreign funds. In particular these changes would be expected to operate most strongly in the case of capacity oriented to the local market. They could more readily overcome such obstacles as limited capital or exchange funds and the modest scale of their own resources and still hope to have a reasonably efficient, though small-scale, industry.

This still does not answer the question of whether the major firms serving industrial countries would be prompted by adoption of the new processes to move nearer to the ore sources. Although transport costs seemingly would favor this, the margin might prove small. Investment exposure would be reduced but would remain substantial. The companies, thus, probably would remain reluctant to make the shift in the absence of greater evidence of stability. At the same time the pressure for smelters from raw-materials-producing countries would mount if they saw themselves deprived of their alumina plants, and this factor could be of some significance.[21]

In summary, the prospective new technology will be likely to favor smaller-scale market-oriented industries in less-developed countries. It is not apt to bring a wholesale shift of capacity to serve industrial countries into such less-developed areas, but both economic and political consequences of the new technology will mildly favor such migration. Again, whether such migration occurs will hinge on questions of investment security and the extent of any power-cost advantage to be had in the less-developed countries.

[21] Australia would face the greatest problem because of the vast distance to market for its bauxite. However Guinea and Guyana undoubtedly would press strongly for smelters in this case.

CHAPTER NINE

FUTURE SOURCES OF ELECTRIC POWER
FOR ALUMINUM SMELTERS

The availability of low-cost electric power traditionally has been a key factor in the siting of aluminum smelters. Niagara Falls, the Swiss and French Alps, and Bavaria were the early centers of the industry. Major production also has grown up in Canada, Norway, and the American Northwest, although in none of these instances are other raw materials found nearby nor is there an important local market for the product.

As power demand has grown in industrial countries, the aluminum industry has had to face the prospect of more expensive electricity in some countries where it wished to expand on a domestic basis. For years it has been recognized that power could be generated at comparable or lower costs in tropical areas which often are nearer to sources of bauxite, yet there has been no rush to build at such sites. Evidently power cost has not been an overriding consideration, and a considerable range of power costs has been tolerated in building new plants. This is explained in part by national policies and by trade barriers which frequently apply to any trade in metals. Other cost factors, plus the added hazard of operating in what is often a less hospitable institutional environment, have dampened enthusiasm on the part of the international companies for locations in nonindustrial countries.

Despite the importance of these other factors, power remains an important cost element which still varies with location. In this chapter the future demands for power for use in reduction will be examined along with potential sources and costs. From this examination we can gain a clearer idea of the alternatives open to the industry and the extent of the cost penalty which may be incurred if the least cost solution is not selected.

ESTIMATION OF FUTURE POWER DEMAND

In the discussion of future aluminum consumption an estimated figure was offered along with a suggestion of the proportion of future consumption which might be supplied by secondary metal.[1] While these figures seem plausible enough and will be employed for the purpose of estimating power demand, other forecasts of demand calling for different primary capacity are easily incorporated into the analysis.

It will be recalled that present typical operating practice uses 17,000–17,500 kwh per metric ton of metal from smelters while under best practice figures as low as 14,000–15,000 kwh have been cited.[2] Moreover, prospective technical improvements may cut these figures still further during the period of concern. However, extreme economies in power use are possible only with increased capital costs, and if power is inexpensive it will be used more freely. For the period as a whole, a figure of 15,000 kwh per metric ton of metal appears a reasonable average for new capacity.

Generating capacity to serve a smelter must provide not only this amount of power but must also allow for transmission and conversion losses, for scheduled maintenance of generating plant, and for unscheduled outages. The reserve capacity needed for these purposes is not easily reduced to a formula. In the past, as a rule of thumb, total generating capacity of about 2¼ kw per metric ton of smelter capacity could be used, but increased efficiency reduced this figure. At 15,000 kwh per ton and allowing for conversion and step-down losses of 5 per cent, about 1.8 kw of generating capacity is needed to support a ton of smelter capacity.

In addition, scheduled maintenance and unscheduled outages probably would not permit a steam electric plant to operate at much more than 90 per cent of capacity on a continuous basis. Assuming ample water supply, a hydro plant might do better, but a sustained figure of 90 per cent of capacity implies that reserve generating capacity must be available either directly or through an intertie. For convenience, this will be assumed to be 10 per cent of the previous total. Thus on an industry basis, and assuming the

[1] Cf. Chapter 3, p. 73ff.
[2] United Nations Secretariat, *Pre-investment Data on the Aluminum Industry*, ST/ECLA/Conf.11/L.24, January 28, 1963, by Jan H. Reimers, pp. 29-30.

availability of interties with electrical grids, a requirement of about 2 kw of generating capacity per ton of smelter capacity can be used as an approximation.

Finally, it cannot be expected that smelter capacity will be maintained at the precise level of consumption. The industry will prefer some margin to meet peak demands and to stay slightly ahead of the growth of consumption. A reasonable target figure may be for smelter capacity to exceed consumption by about 15 per cent. In practice this margin will fluctuate and it may go higher than this as firms jostle for position in the market.

On this basis, added power demands exclusive of Communist-bloc countries can be computed at 19.0 million kw of generating capacity by 1980. This is derived in the following manner:

Derivation of a Forecast of Power Demand for Smelting

	1980 (million metric tons)
Total expected aluminum consumption	17.5
Less estimated secondary consumption	4.6
Equals total primary consumption	12.9
Plus allowance for excess capacity (15%)	2.3
Equals total primary capacity desired	15.2
Less current (1965) capacity	5.7
Equals net added capacity required	9.5
Net added generating capacity at 2 kw per metric ton	19.0 million kw

Obviously, if generating capacity is located at points distant from smelters or if it includes power from less reliable hydro sites, some increase would be required to assure sufficient power. However, more likely sources of error in these figures are in the estimates of consumption, above all, and in the assumptions concerning future power inputs per unit of production.

SIGNIFICANCE OF POWER COST AND EFFECT ON LOCATION

At given ratios of power use per unit of production it is a simple matter to indicate the significance of variations in power cost. In practice, power will be used more carefully when more expensive, and vice versa, yet there are compensating costs and savings in these circumstances, so it is not inappropriate to assume an average power input for the purpose of calculating costs.

177

At 15,000 kwh per metric ton and with most power costs about the world ranging over a span of from 2–8 mills per kwh, the power cost could vary from $30–$120 per metric ton.[3] A difference of one mill in power cost will be roughly the equivalent of $15 per ton difference in cost. The relative importance of this can be discerned from the fact that metal currently (1965) sells for $540 per metric ton on the world market. Thus a difference in power cost of one mill is equivalent to roughly 3 per cent of selling price or a somewhat larger proportion of cost.[4] Provided that it is not offset by other costs, a difference of 3–4 per cent is not of negligible importance; a difference of several mills can be quite significant.

As was noted, a spread in power cost between the highest and lowest costs amounting to $90 per ton on metal selling at $540 per metric ton exists under present conditions. It may be tolerated because of compensating differences in other costs or because of trade barriers protecting high-cost production. For example, note that, at present prices and with the anticipated rate of power consumption, a tariff of 17 per cent would permit even the most unfavorably situated domestic producer to compensate for his entire disadvantage in cost of power vis-à-vis his most favorably situated foreign competitor. Thus, while favorable power costs exert a major locational pull they are not so compelling as to be able to override the influence of moderate tariff levels or of disadvantages in other components of cost. By the same token, a country determined to produce its own metal ordinarily need not face a wholly forbidding penalty because of higher power costs.

POWER-MARKET SITUATION AND DISLOCATION OF EXISTING SMELTERS

Existing aluminum smelters often enjoy exceptionally inexpensive electricity, sometimes at costs far below what the power would bring if devoted to other uses. The reasons for this are varied and are rooted in the history of the industry. However, since the alumi-

[3] Bache and Co., *The North American Aluminum Industry* (New York, March 1964).
[4] The Senate Rampart Power study figures cost ex profit in the United States at about $340 per short ton (equals $375 per metric ton). Under these conditions, 1 mill would amount to 4 per cent of cost. U.S. Senate Committee on Public Works, *The Market for Rampart Power, Yukon River, Alaska*, 87th Cong., 2nd sess. (Washington: U.S. Government Printing Office, 1962), p. 155.

num industry has no exclusive claim on low-cost power sources while other industries pay more, it might be expected that as electricity consumption in developed countries increases competing demands for electricity would divert power to other uses. As we shall see, diversion of existing power capacity is apt to be less of a problem than it appears to be at first sight.

To the extent that companies buy power on long-term contracts, diversion is delayed by the existence of such contracts. Even if the price charged for power were to rise sharply over time, aluminum companies might prefer to pay it rather than abandon their facilities. While the higher price might be damaging to the earning capacity of existing investment, it need not result in closing the plant and diverting significant amounts of power.

In the United States and Canada a major share of the power used by the industry also is owned by aluminum firms. They would not divert this to other use except at very improbable prices for electricity, for it would entail abandoning existing investment in reduction plants. The picture for other areas is mixed, with electricity sometimes purchased and in other cases produced internally. On occasion (in France, for example) the power is provided by state enterprises and proposed diversion to other uses could be opposed through political channels by the affected companies, with the outcome slow and uncertain by contrast to what would be expected in response to purely market processes. In the United States, it seems very improbable that those government agencies now providing power to a portion of the aluminum industry (TVA, Bonneville, and New York State Power Authority) will establish power prices which are so burdensome to the industry as to compel its relocation during the period of concern. Far more likely is the possibility that they will discourage new plant locations premised on exceptionally cheap power, although even here there have been recent exceptions.

Up to the present, aluminum reduction has favored sites with lower power costs and often has been willing to go to isolated locations to obtain them. Where it occurs, this very isolation has provided a kind of protection against diversion of power to other uses. However, the prospect of more complete electrical grids with long-range high-voltage transmission lines capable of moving power economically to higher-price markets reduces the security to be found in such isolation. Increasingly it becomes feasible to market elsewhere the power from remote sites. A concomitant development, however, may diminish the likelihood of such diversion: namely, the ever-greater possibility that grids can be fed with low-cost thermal power from more advantageously sited hydrocarbon-fueled or nuclear-fueled plants. Indeed, in the United States the aluminum industry relies to a major extent on thermal power whose cost is stable or declining and which is virtually unlimited in supply. In this circumstance there is no threat of diversion to higher use. Thus, the spread between low-cost isolated hydropower delivered to load centers and the more conveniently available thermal power is diminished, and with it the pressure for diversion.

For all of these reasons the likelihood of diversion of power from existing aluminum plants to other uses is not very great. The possibility is greatest in those (rare) instances where hydropower is purchased from private suppliers. In such cases aluminum plants would still find it advantageous to pay higher power prices so long as operating costs for metal produced from these sites remains below full costs from alternative sites. Theoretically, an analogous concept might apply for diversion of their owned power supply; they would be best advised, that is, to sell the power and dismantle the plant at this point because their returns on power sold would be high enough to justify it. In practice such alacrity is uncommon. Where power is purchased from a government agency diversion from existing plants seems a very remote possibility. Thus, the increase in power demand, if reflected at all in higher power costs, is more apt to influence new locations than to result in a shift of existing capacity.

CHARACTERISTICS OF CONVENTIONAL THERMAL POWER

At present, conventional thermal power is widely used in the aluminum industry in the United States and Europe. While the most economical thermal power cannot attain the low costs of the cheapest remaining hydropower on a world basis, within some developed countries it often proves cheaper than remaining domestic hydro sites, or there may be advantages of location which offset the higher energy costs. Moreover, even in less-developed

180

countries very inexpensive fuel sources are available in some instances which could be used to generate thermal power at prices which might be attractive to the industry—the most likely case being where otherwise unmarketable natural gas is available as a by-product of petroleum production. However, it is necessary to look at the costs of conventional thermal power primarily because of its continuing importance to the aluminum industry in developed countries.

Generation of thermal power has been characterized by a steady trend toward large plants and units which has restrained unit capital cost and has brought improved thermal efficiency with attendant favorable effects on operating costs. In addition, advances in the extraction and transportation of mineral fuels have held fuel costs, especially in parts of the United States, to levels which make thermal power attractive to aluminum producers.

A large aluminum smelter of 200,000-metric-ton capacity would require the entire output of a generating plant in the 400-mw range or a portion of the output of larger plants. Efficient generating plants now tend to run to this size or larger.[5] More important has been the trend toward larger units within individual plants where new installations commonly run above 300 mw and go as high as the 700-mw range for a single turbine generator.[6] Aluminum companies will be reluctant both to forego the efficiency of the larger units and to allow their entire operation to depend on a single unit. In consequence they may be expected to seek arrangements with existing utilities wherever possible so as to have back-up power available either through cooperative ownership of generating capacity or through standby tie-ins with existing grids. Although in some cases there may be legal or competitive obstacles to be overcome, ordinarily such arrangements will be possible and the industry should be able to take full advantage of the economies to scale occurring in steam generation.

An AEC study anticipated that, in the United States, conventional coal-burning steam-generating plants could be built for

[5] Note, for example, that the 10 largest steam plants in the United States ranged from 1,600 mw to 1,155 mw, as reported in Federal Power Commission, *Steam-Electric Plant Construction Cost and Annual Production Expenses,* 16th Annual Supplement (Washington: U.S. Government Printing Office, 1963), p. xxiii.

[6] *Ibid.,* p. v.

operation by 1966 at about $129 per kw for plants of 500 mw. Similar plants were figured at $110 per kw by 1980.[7] Costs of recent plants have varied considerably, depending on size, type of fuel use, and whether they are enclosed or outdoor plants. However, *Electrical World's* "14th Steam Station Cost Survey"[8] showed that new plants or new units which were installed in the United States in 1962–64 averaged $127 per kw with large (300–1,000 mw+) coal-fired stations generally over $140 per kw. Gas- and oil-fired stations were significantly cheaper. An authoritative industry source has estimated capital costs for an advanced design coal-fired plant of 800 mw at $112 per kw in 1966.[9]

The significance of such capital costs for energy costs depends on the plant factor and the fixed charges assumed. Utilities normally assume a high plant factor during the early years of a plant's life, declining thereafter as newer capacity with lower operating costs takes over the base load. Since the demand for energy from a smelter is virtually constant, a generating plant built to serve this industry could be expected to operate at a very high plant factor throughout its life.[10] Utilities often figure on plant factors as high as 80 per cent, and where virtually the only fluctuation stems from maintenance needs rather than demand variations, still higher factors may apply.[11] Scheduled maintenance will take generating capacity out of operation for 6–8 per cent of the time and unscheduled outages, increasing with the age of the plant, will add to this. Thus, lifetime plant factors in the 80–90 per cent range appear reasonable for capacity serving the aluminum industry.

Fixed charges consist of the cost of money (whether borrowed

[7] U.S. Atomic Energy Commission, *Civilian Nuclear Power—Report to the President, 1962* (Washington, 1962), p. 59 of Appendix.

[8] *Electrical World*, October 18, 1965, pp. 107-8.

[9] Talk by Philip Sporn, "Nuclear Power Economics," New York, March 17, 1966.

[10] This would be the case if the plant were owned by the aluminum company. A utility providing power to the smelter might divert the generator to peak-load operation at some point, but a bloc of steady demand improves the system load factor and, from a system standpoint, they still would have an incentive to price the power to the smelter on what is the equivalent of an assumption of a high-lifetime plant factor.

[11] The Oyster Creek nuclear station expects an 88 per cent plant factor during the first 15 years for a plant operating as part of a system. See Jersey Central Power and Light Co., *Report on Economic Analysis for Oyster Creek Nuclear Electric Generating Station*, February 17, 1964.

or equity), depreciation, insurance, and taxes. Total fixed charges per year are divided by total output of energy to arrive at their contribution to the cost of energy. Since the total does not vary significantly with output, fixed charges per unit of output are lower at higher plant factors. Because of the importance of depreciation and the cost of money in total fixed charges, the level of fixed charges is most affected by the size of the investment outlay per unit of output and by the rate paid on investment funds.

In making their calculations for American conditions, the AEC assumed annual fixed charges of 14 per cent of the capital cost of the plant.[12] A breakdown of typical American fixed charges totalling 14.54 per cent provided by Teitelbaum shows that 5.85 per cent is taxes and 0.80 per cent insurance, so that cost of money, depreciation, and inventory come to only 7.89 per cent per year.[13] The American breakdown was based on a mixture of 50 per cent borrowed funds, 15 per cent preferred, and 35 per cent common stock with appropriate rates for each. Investment in other areas could expect to find different tax charges, perhaps lower, but frequently would encounter higher costs of money and shorter depreciation periods with attendant higher costs. Commercial money rates would be higher in most other countries. Of course, governments frequently could borrow at or below the 6.35 per cent weighted average used in Teitelbaum's example. However, as he notes, the cost of domestic funds diverted through government borrowing is the yield which the same funds would have earned in some other uses.[14] Properly calculated on this basis, fixed charges may be assumed to exceed 10 per cent in nearly all circumstances and may be far more where risk is high and depreciation period short. The implications for energy costs of various assumptions concerning plant cost, plant factor, and fixed charge rate are shown in Table 18.

The principal operating cost of thermal power is for fuel. Fuel prices vary greatly from place to place even within a country, so it is difficult to generalize about them. Operating costs also are affected by the thermal efficiency of the plant and by costs of repair, maintenance, and labor and supervision.

[12] U.S. Atomic Energy Commission, *op. cit.*, p. 55 of Appendix.
[13] Perry D. Teitelbaum, *Energy Cost Comparisons*, Reprint No. 38, Resources for the Future, Inc., 1963.
[14] *Ibid.*, p. 212. This statement should be modified to allow for differences in financial risk.

Table 18. Effect of Changes in Unit Capital Cost, Fixed Charge,
and Plant Factor on Energy Cost

A

Effect of One Percentage Point Variation in Fixed Charge on
Cost of Energy in Mills per Kwh at Various Plant Factors

Capital Cost of Plant	Plant Factor		
	80%	90%	95%
$100	.143	.127	.120
110	.157	.140	.132
120	.171	.152	.144
130	.186	.166	.156
140	.200	.179	.168
150	.214	.192	.180

B

Illustrative Table of Fixed Charges in Mills per Kwh Under
Differing Assumptions Concerning Unit Capital Costs,
Plant Factors, and Fixed Charge

Capital Cost of Plant and Fixed Charge Rate	Plant Factor		
	80%	90%	95%
$100			
At 10% fixed charge	1.43	1.27	1.20
At 15% fixed charge	2.14	1.90	1.80
$120			
At 10% fixed charge	1.71	1.52	1.44
At 15% fixed charge	2.57	2.28	2.16
$150			
At 10% fixed charge	2.14	1.92	1.80
At 15% fixed charge	3.21	2.88	2.70

The AEC study assumes operating costs exclusive of fuel in future coal plants to be about 0.27 mills per kwh, a figure which is well below the industry average but not out of line with the 0.45 average attained by the TVA coal-fired plants in 1961, or the still lower rates of individual plants reported by FPC.[15] Most of the operating costs are for labor, and they run higher for coal plants with their bulk fuel and ash problems than for gas- or oil-fired plants.[16] They could be expected to be still higher for plants burning lignite because of the increased bulk of material handled.

[15] U.S. Atomic Energy Commission, *op. cit.*, p. 55 of Appendix; Federal Power Commission, *op. cit.*, 14th Annual Supplement (1961), p. xxx.
[16] *Ibid.*, p. xiv.

Fuel cost per kwh depends upon unit costs of fuel and on thermal efficiency. At a given cost for fuel, improved thermal efficiency brings proportional cost savings (Table 19). Long-term

Table 19. Illustrative Table Showing Relation of Fuel Costs and Thermal Efficiency in Mills per Kwh

Cost of Fuel in Cents per Million Btu	Thermal Efficiency in Btu per Kwh			
	10,000	9,000	8,000	7,000
	(fuel cost in mills per kwh)			
1	0.1	0.09	0.08	0.07
10	1.0	0.9	0.8	0.7
15	1.5	1.35	1.2	1.05
20	2.0	1.80	1.6	1.40
25	2.5	2.25	2.0	1.75
30	3.0	2.70	2.4	2.10
35	3.5	3.15	2.8	2.45
40	4.0	3.60	3.2	2.80

improvements in thermal efficiency have been dramatic and their effect on the cost of electrical energy has been important. Now, however, it is thought that there are limits to future improvements in thermal efficiency and the rate of improvement will slow. The best American heat rate reported in 1963 was 8,646 Btu per kwh for a coal-fired unit.[17] It has been speculated that new plants may attain 7,500 Btu per kwh by 1980, and the AEC calculations for 1966 were based on a rate of about 8,500 Btu per kwh.[18] As a compromise, a rate of 8,000 Btu by 1980 may be taken as a conservative standard.[19]

At a rate of 8,000 Btu per kwh, fuel costs in mills per kwh can be computed based on varying prices of fuel, with each cent of fuel cost per million Btu equal to 0.08 mills in kwh cost (Table 19). Thus, fuel priced at 25 cents per million Btu would contribute 2

[17] Federal Power Commission, op. cit., 16th Annual Supplement (1963), p. xxviii.
[18] For the 1980 figures see H. H. Landsberg, L. L. Fischman, and J. L. Fisher, Resources in America's Future (Baltimore: Johns Hopkins Press for Resources for the Future, Inc., 1963), p. 288; U.S. Atomic Energy Commission, op. cit., p. 55 of Appendix.
[19] This agrees with the figure suggested by the Federal Power Commission, National Power Survey, Part II—Advisory Reports (Washington: U.S. Government Printing Office, 1964), p. 347.

mills per kwh to the cost of energy. As a rough approximation, this is the equivalent of coal at $7.20 per metric ton. Fuel cost must be figured on a delivered basis, of course, but thermal plants to serve aluminum smelters often may be located near the coal fields. It is interesting to note that the 1963 average mine price of coal in the United States was $4.39 per short ton (= $4.83 per metric ton) or the equivalent of about 18 cents per million Btu, which in turn implies 1.5 mills per kwh at the prospective heat rate envisioned.[20] In four states as of 1962, steam electric plants burned coal at an average cost of under 20 cents per m/Btu, the lowest being 17.3 cents in Kentucky. Individual plants burned coal as low as 14–16 cents per m/Btu in the Ohio Valley.[21]

These two principal components of cost—fixed charges and operating costs—may then be pulled together in an illustrative manner under the following assumptions for 1980: (1) fixed charges of 15 per cent, (2) capital cost of $110 per kw, and (3) 90 per cent plant factor, which combination implies 2.1 mills per kwh fixed charge, and (1) fuel costs of 20 cents per m/Btu, (2) 8,000 Btu per kwh heat rate (this combination equal to fuel costs of 1.6 mills per kwh), and (3) other operating costs of 0.3 mills for 1.9 mills per kwh operating costs or a total cost of 4 mills per kwh. Such total costs are equalled or bettered by some existing American plants, although in few other areas is there so favorable a juxtaposition of low money rates, moderate capital cost, and inexpensive fuel.

Since it is based on American operating experience, the foregoing discussion of thermal power has particular application to the United States. However, the technology employed in Europe is similar, although units tend to be smaller. This technology is transferrable to more remote areas, the more so if the power plant is to be built by major international firms knowledgeable in this field. Capital costs per unit of capacity may be higher in nonindustrial countries because of the need to import equipment and skilled

[20] U.S. Bureau of Mines, *Minerals Yearbook, 1963* (Washington: U.S. Government Printing Office, 1964), Vol. II—Fuels, p. 162. At prevailing thermal efficiencies and fuel costs, two major American power *systems* had fuel costs of 1.69 and 1.78 mills per kwh in 1963, both using coal (Federal Power Commission, *Steam-Electric Plant Construction Cost, op. cit.,* 16th Annual Supplement (1963), p. xxx).

[21] National Coal Association, *Steam-Electric Plant Factors, 1963* (Washington, 1964), p. 22.

construction personnel. Likewise, commercial money rates may be higher in such countries, whether dependent on domestic or foreign funds. There is scant possibility of saving on operating costs other than fuel because such costs already are so small and nonindustrial countries frequently lack highly skilled operating and maintenance personnel.

Thermal power inevitably entails fixed charges of upwards of 2 mills per kwh. If it is to be employed in nonindustrial countries for aluminum reduction, it must have access to very inexpensive fuel as a minimum and probably also a favorable juxtaposition of raw materials and transport facilities. A simple example will illustrate this point. Suppose a plant built at $110 per kw in the United States costs $130 (or about 18 per cent more) set down in a remote area. Even if fixed charges for the two plants are the same at 15 per cent per year and plant factors are constant at 90 per cent (both very generous assumptions), fixed cost will be 0.4 mill more in the nonindustrial area. This is the equivalent of 5 cents per m/Btu at the assumed future heat rate. Since the most economical coal plants in the United States get fuel for around 15 cents per m/Btu, the remote plant would have to get fuel at 10 cents per m/Btu, or less, simply to match U.S. power costs. In practice, the most likely competition would be from gas-fired thermal stations whose U.S. costs per kw of capacity are significantly lower than coal stations—perhaps $20–$30 less in the relevant size range.[22] This might offset the added construction costs of a remote area and give full weight to its fuel cost advantage. It is noteworthy that the most abundant source of inexpensive fuel, i.e., Middle East gas, has been considered as a possible location for aluminum smelting facilities, but so far none has been constructed there. Remoteness from raw materials and markets and other factors may diminish the attractiveness of low fuel costs for electric generation.

PROSPECTS FOR CONVENTIONAL FUELS

Focussing for the moment only on fuel costs for thermal generating plants, where are abundant and inexpensive fuels to be found? The question is not as simple as it may appear, for our concern is not only with the period out to 1980 but also beyond that

[22] *Electrical World,* October 7, 1963, p. 78.

for the time required to amortize investments to be made as late as 1980. It can be anticipated that on a world basis many fuel sources presently unknown will have been discovered prior to 1980 and that rates of exhaustion of known reserves will alter in a not wholly predictable fashion. Future costs of extraction of known reserves are uncertain, while not even guesses can be made concerning extraction costs and location of unknown reserves. Aside from costs of extraction, national or company policies will affect prices of these fuels so that price may diverge sharply from cost.

Obviously, we cannot deal with all these intangibles here. It seems best to take an inventory of those sources of fuels which are accessible, abundant enough to be of long range interest, and currently priced at a level which might be considered by the industry. This information can be viewed in the light of developments in extraction and transportation in the United States which may affect costs elsewhere.

The United States has abundant and inexpensive fuels at sites accessible by barge from ocean ports. This means that the raw materials of the aluminum industry can be transported by water to smelters adjacent to mine-mouth generating plants for reduction. The importance of this is apparent when it is realized that 4–5 metric tons of coal are required for thermal plants to produce the 15,000 kwh required per ton of metal while other raw materials amount to only about 2.5 tons.[23]

Recoverable coal reserves in the United States are variously estimated, but one figure based on 50 per cent recovery places the total at 830 billion tons—a supply sufficient to carry the United States far beyond the period of interest here.[24] With reserves of this magnitude there is no question about the continuing availability of coal-fired generating potential for aluminum production. This conclusion remains true even after considering all other prospective uses for coal during the period of interest.

Likewise, on costs the picture is very optimistic for the United States. No one can be certain what future fuel costs will be since they depend on the interplay of changes in wage rates, equipment

[23] See footnote 25, Chapter 7; other materials are alumina (2 tons) and carbon (0.5 tons). Reimers, ST/ECLA/Conf.11/L.24, *op. cit.*, p. 29.

[24] U.S. Geological Survey *Coal Reserves of the United States—A Progress Report, January 1, 1960*, Bulletin 1136, by Paul Averitt (Washington: U.S. Government Printing Office, 1961), p. 1.

prices, technology, and resource quality. However, so far as the quality of resources is a determinant, it appears that the United States has ample reserves of coal of such quality that under current price relationships they could be mined at present prices. The Federal Power Commission study suggests that for the 1970–80 period coal may be produced at the mine at a price of about 13.5 cents per million Btu.[25] Since ample amounts of this coal are found in areas served by Mississippi River system barge routes, its continuing relevance for aluminum production is assured.[26]

Although natural gas is used as a fuel to produce low-cost electricity for aluminum reduction in the United States, no further expansion of this practice seems likely because of the growing market for gas and its improved access to market. Likewise, petroleum enjoys higher uses and is not priced so as to permit production of low-cost power, nor is there any real prospect that it would be able to compete with coal for low-cost generation in the United States.

The significance of this brief excursion into the U.S. fossil fuel situation is that, for the contiguous areas of the United States, quite apart from any future developments in atomic energy or any potential for further development of hydro, the American industry can be expanded readily on the basis of coal-fired electric plants at costs which are likely to prove attractive domestically. This does not necessarily mean that the United States could expect to compete in foreign markets with metal produced on this basis, but it does suggest that no great gap is likely to arise between U.S. consumption and productive capability which would need to be filled by foreign metal.

Outside of the United States it is possible to indicate in a general way where major reserves of fuels occur by referring to other series compiled by the World Power Conference.[27] Referring only to measured reserves, the countries with coal supplies on a very sizable scale include the United States, Germany, Canada, South Africa, and the United Kingdom among non-Communist countries,

[25] Federal Power Commission, *National Power Survey, Part II, op. cit.,* Advisory Committee Report No. 21, p. 322.

[26] U.S. Department of the Interior, *Supplies, Costs and Uses of the Fossil Fuels,* by the Energy Policy Staff (Washington, February 1963), Table 2 (mimeo).

[27] World Power Conference, *World Power Conference Survey of Energy Resources, 1962* (London, 1962).

and U.S.S.R., Poland, and presumably China in the Communist bloc.[28] In relation to the size of the domestic economies and competing needs for fuel, considerable reserves also are available in Australia and the Netherlands (Table 20).[29] Brown coal and lignite are found in Germany, Australia, Canada, and the United States, as well as the U.S.S.R., while the most extensive inferred deposits are concentrated especially in the U.S.S.R., United States, and Australia.[30]

Estimating reserves of oil and gas is trickier still, since the presence of oil is harder to discern. Staying with the estimates of proved reserves, the truly large deposits are found in the Middle East, United States, U.S.S.R., Venezuela, North Africa, and Indonesia, while Canada, Mexico, and Argentina also have sizable deposits. Natural gas supplies have been less thoroughly explored, with the United States predominating in known supply. The U.S.S.R., Netherlands, Algeria, Venezuela, Canada, and Pakistan are known to have substantial reserves, and the Middle East undoubtedly is also well endowed.[31]

Availability is not merely a question of the size of fuel reserves but also of anticipated levels of consumption. Long-term consumption projections are available only for some areas. While this might appear to be a serious problem, it proves in fact less troublesome so far as oil is concerned. Oil moves readily in trade at low transport cost, and availability is not a function of domestic reserves. Gas does not move as easily, and only tidewater or readily accessible gas fields would provide really attractive cost prospects to the aluminum industry. At prices of interest to the aluminum industry, coal also tends to be place-bound and the number of suitable deposits is limited.

[28] This concept is used because a commercial venture would require this kind of certainty.

[29] Purely as an illustration, we may assume that at least a 30-year supply of fuel must be available to justify the location of a coal-fired thermal electric plant. At 15,000 kwh per metric ton and 8,000 Btu per kwh and 28 million Btu per metric ton of coal, this implies consumption of approximately 430,000 tons of coal per year or 13 million tons over 30 years for a smelter of 100,000-ton annual capacity. Since nearly all other uses for coal outbid the aluminum industry, reserves must be of such magnitude in relation to demand that the required amount of coal will be available to the aluminum industry at lowest cost. In most circumstances this implies reserves of many times the anticipated consumption for smelting.

[30] World Power Conference, *op. cit.*, pp. 23-25.

[31] *Ibid.*, pp. 31-37; also *World Oil*, August 15, 1964 (Houston), p. 95.

Table 20. Reserves of Coal in Selected Countries
(million metric tons)

Country	Measured Reserves	Total Reserves Including Indicated and Inferred
Canada	42,000	60,000
United States	72,000	1,100,000
Colombia	—	12,500
Argentina	374	454
Brazil	388	1,700
Belgium	1,557	3,425
France	U[a]	4,400
Germany, Federal Republic of	70,200	229,900
Italy	675	700
Netherlands	2,394	2,394
Spain	2,784	2,957
United Kingdom	127,079	169,379
Turkey	489	1,298
Japan	5,723	19,248
India	—	57,832
Australia	1,791	12,844
Botswana	506	U
Southern Rhodesia, Zambia, and Malawi	1,760	6,613
South Africa, Republic of	21,443	63,355
Swaziland	2,022	5,022
Tanzania	300	400
Czechoslovakia	—	6,450
Poland	75,500	135,000
U.S.S.R.	143,688	4,630,050

[a] U = Unknown
Source: World Power Conference, *World Power Conference Survey of Energy Resources, 1962* (London, 1962), Table II, pp. 20-22. See this source for basis of estimates.

It is pointless to try to analyze gas availability in detail because exploration for gas is only now beginning to attract serious attention apart from oil, and techniques for moving it to higher use in distant markets may yet be perfected. Domestic aluminum industries based on gas now exist in the United States and France, and a plant is planned for the Netherlands. Except (possibly) for the Netherlands, it is unlikely that these will be expanded because

191

of better use for the gas. Other such plants on a small scale might be feasible for local use in Argentina and Venezuela. Under present conditions of known availability, the only really large-scale possibilities for the use of gas would be in the Middle East or North Africa, where other uses are less likely to bid up the price of the abundant local supply. Since the Middle Eastern gas is a by-product of oil production, its price, apart from costs of collection and storage, is arbitrary. Considerations of distance from materials and markets, climate conditions, and the uncertain political situation of the Middle East all diminish the attractiveness of the area. Certainly an extremely low gas price (surely under 10 cents per m/Btu) would be needed to interest the aluminum industry, and given present alternatives, even a nil fuel cost might not suffice.

Of those countries mentioned as having sizable reserves of coal, only the United States, South Africa, and Australia appear to have coal recoverable at a price and location of interest to the aluminum industry. German coal, like other European coal, is expensive, in the range of $15 or more (or about 50 cents per m/Btu). British coal is cheaper in some locations but still expensive at $10 or so per ton. The Japanese also find coal expensive, of poor quality, and full output is needed for other purposes.[32] South Africa produces some of the world's least expensive coal, with prices as low as $2 or less at pit head. While this is inland and not favorably situated relative to known bauxite sources, it would permit a domestic industry based on nonbauxitic materials as a hedge against possible trade boycotts. It could not be viewed as a probable export base, since moving either coal or bauxite by surface transportation would quickly raise costs. Australia presents the more interesting case in view of its enormous supplies of bauxite. Prices of New South Wales coal range in the $5–$6 class located near the sea. Australia is anxious to process its bauxite raw materials as fully as possible. Since the country is short of hydropower, a coal-based aluminum industry cannot be excluded; however, the distance to market is such that, added to the implied thermal power costs, it is not likely to result in very rapid expansion of the Australian production for export.

Lignite resources also can be used for power production in Australia (one plant already draws power from the Victoria power

[32] Based on discussion with U.S. Bureau of Mines.

grid heavily based on lignite), but it is doubtful that this will prove more economical than the better-located coal of Australia's east coast. Germany has considerable lignite reserves, some of which are used to produce power for the aluminum industry. This is expandable, apparently at costs of around 30 cents per m/Btu.[33] The calorific value of Rhineland lignite is about one-fourth that of coal, so even the very cheapest open-pit mining techniques cannot produce equivalent amounts of heat at prices lower than coal in the United States. However, the German lignite very well may produce future electricity in the 5-mill range.[34] As the supply of low-cost lignite is limited, rational use of this resource would require that it be devoted to general industrial needs rather than reserved for aluminum production.

Whether a given cost of fuel will be attractive for the aluminum industry depends very much upon local conditions and government policies. Assuming that the abundant coal reserves of the United States will provide virtually unlimited thermal power at a fuel cost of under 20 cents per m/Btu and an implied power cost of about 4 mills, a kind of ceiling on cost is established. Higher fuel and power costs may be justified elsewhere but only if better logistics for the aluminum industry are obtainable or if trade conditions permit. Where export industries are concerned and special trading blocs are not in question, logistics can be an important factor. Conceivably, areas located near bauxite reserves under certain circumstances might tolerate fuel costs higher than the United States and still be able to compete with its low-cost thermal-based production in third-country markets. Australia may provide a test of this possibility in Far Eastern trade. Again, if a country pursues a policy of self-sufficiency it may tolerate fuel and power costs far above those required to make its industry freely competitive with external sources of metal.

So far as conventional thermal power is concerned, costs of 20

[33] Arbeitsgemeinschaft deutscher wirtschaftwissenschaftlicher Forschungsinstitut (Bonn), *Untersuchung über die Entwicklung der gegenwärtigen Structur von Angebot und Nachfrage in der Energiewirtschaft der Bundesrepublik unter besonderer Berücksichtigung des Steinkohlerbergbaus* (Berlin: Duncker and Humblot, 1962), pp. 134-6. The short title of this document is *Energie-Gutachten, 1961.*

[34] *Metal Bulletin*, March 17, 1964 (London), cites a current figure as low as 3.5 mills. This implies exceptionally low mining costs. A higher figure for future power is suggested in *Energie-Gutachten, op. cit.*, p. A215, but this is based on a very low plant factor appropriate for a utility system.

cents or under per m/Btu implies coal at less than $5.75 per metric ton, oil at less than $8.40 per metric ton, or gas at under $7.60 per 1,000 cu/m.[35] As was suggested, only the United States and Australia are known to have substantial available coal in this price range so located as to be accessible to water transportation. South Africa could support a domestic industry based on coal and inferior domestic ores if determined to do so for policy reasons.

Quoted prices for oil at major world production centers substantially exceed the 20-cent figure, but these prices often are deceptive. Depending on grade, oil would need to sell for about $8.40 per metric ton or $1.20 per barrel to attain the 20 cent figure per m/Btu. In fact, delivered oil prices are well above this range and oil does not appear likely to support an export aluminum industry.[36] However, if a country wishes to pursue a policy of self-sufficiency in aluminum and lacks adequate coal or hydro, oil offers an alternative power source. This would seem of most immediate interest to Japan where coal is in limited supply and very expensive. Here oil, even at 35 cents per m/Btu (or about $2 per barrel, or $13 per metric ton) can provide a less expensive source of power than domestic coal.[37] It is particularly difficult to make definitive statements concerning prospective oil prices because in individual circumstances production costs may be far lower than prices would indicate and quoted prices frequently are shaded by a significant margin.[38]

From the foregoing it is apparent that, to the extent that market forces alone govern, significant expansion of aluminum smelting

[35] Calorific values of these fuels vary, of course, but the figures cited are based on the following conversion factors, i.e., 28.7 million Btu per ton of coal, 41.9 million Btu per ton of oil, 38.0 million Btu per 1,000 cu/m of gas.

[36] See *Petroleum Press Service* (London) for prices. Prices vary, but $2 per barrel or $14 per ton is a common figure. Informed sources claim that oil is now laid down in northwest Europe at $11-$12 per metric ton, but local taxes must be added in some cases.

[37] The same would be true for most of Europe if it should wish to expand conventional thermal-based aluminum production. Note oil already is used widely in Italy for production of electricity.

[38] It has been argued that the production cost of Middle East oil is only a tiny fraction of price up to this point and that if price should come to reflect cost, oil could be delivered to Europe for not over 11 cents per m/Btu (M. A. Adelman, *Oil Production Costs in Four Areas*, American Institute of Mining, Metallurgical and Petroleum Engineers, 1966). Obviously this would make oil very competitive should national policy permit its entry at such prices. We cannot base our analysis on the assumption that this will come to pass.

based on conventional thermal power is unlikely to occur outside of the United States and possibly Australia. This does not represent any final judgment, of course. For example, the heavy cost penalty paid by Europe and Japan through utilization of domestic coal could be reduced by greater use of imported fuels, provided trade restrictions be modified to allow such fuels to be supplied at lower costs. The Italians and Japanese have shown initiative in cutting the oil price to their economies through special arrangements with producing countries outside the terms of the customary international deal. In other countries, the resistance of coal producers to low-cost oil imports and the pressure of national oil companies to maintain price structures up to now have made this move more difficult. Nonetheless, if countries wish to remain self-sufficient in metal, exceptions could be made to favor the limited amount of energy needed for this purpose.

For our purposes, the chief significance of conventional thermal power is that coal-fired plants in the United States are capable of virtually unlimited expansion to meet the needs of the world's largest market, using well-tested methods, and at costs that are likely to prove tolerable into the future. Moreover, in a context of freer trade in aluminum, this metal will be available to other areas if needed. Thus, to the extent that economics govern, a kind of limit is established on the cost of power from underdeveloped countries which would be attractive to the industry, and the pressure to take that power is diminished. Should nuclear power fully justify the renewed optimism held out for it, this limit may be supplanted in many developed countries, possibly including the United States, by the limit imposed by the cost of nuclear power. In the interim, however, the U.S. coal-fired potential occupies a key position.

PROSPECTS FOR NUCLEAR ENERGY

A search for low-cost power sites assumes that there are differences in power costs as between sites. This is obviously true of conventional sources of power, whether hydro or steam. In the case of hydro the site is dictated by terrain and water availability, while for conventional steam it is largely a matter of the cost of extracting and transporting the fuel to the site. Flexibility in smelter location may be gained as transmission developments permit more economical movement of localized power, thereby giving it greater geographical reach.

However, it is necessary meanwhile to look at the one power source which is par excellence ubiquitous: nuclear power. A short time ago a study such as this confidently could have ignored nuclear energy. Despite a long string of optimistic pronouncements, it had been proved competitive only with very high-cost conventional sources and at prices of no interest to the aluminum industry. Recently, however, this situation has changed and nuclear energy appears more promising. Moreover, the recent cost reductions have been so rapid that they are likely to result in the construction of a number of economically sized production line models which will provide valuable experience in nuclear plant operation and a basis for further advance.

It long has been recognized that the chief advantage of nuclear plants is the low-fuel cost provided at almost any site desired.[39] The chief disadvantage has been the substantially higher capital cost per unit of capacity. Meanwhile improved heat rates of conventional plants and improvements in the extraction and transportation of fuels, as well as in transmission of electricity, have made large, well-located conventional thermal stations still more formidable competitors. Therefore, it was felt that nuclear power would become competitive only with the most expensive conventional power at first and that gradual improvement would bring it within the range of moderately priced conventional power. The significant recent change has been in the capital costs of nuclear plants which, at large sizes, have moved within the range of the most economical conventional plants while retaining an advantage in fuel costs. Thus, it becomes possible to build nuclear plants in developed countries which will produce electrical energy at prices competitive with low-cost conventional thermal power.

Although much of the early optimism concerning the progress of nuclear power proved disappointing, two recent American contracts are of particular interest. An American equipment manufacturer has contracted to build a 515–620-mw boiling water nuclear station with expected energy costs in the range of 3.42–4.49 mills per kwh under commercial conditions of financing, taxes,

[39] Sam H. Schurr and Jacob Marschak, *Economic Aspects of Atomic Power* (Princeton: Princeton University Press for Cowles Commission, 1950), p. 4.

etc.[40] Although the computations involved in this calculation assume what some have criticized as an unrealistic plant factor (88 per cent), the assumed plant factor seems technically feasible and not unrealistic if built to serve a continuous-process industry such as aluminum.[41]

Capital costs for the plant, depending upon the capacity ultimately achieved, fall in the $110–$130 range—little more, if any, than for coal plants of comparable size. At first it was alleged that temporary economic conditions and a "loss-leader" strategy on the part of the manufacturer accounts for this startlingly low price. However, the manufacturer subsequently has posted prices for units of varying size on the assumption that multiple sales will be attained. The prices suggest that future units will be near the same range and competitive with coal at 23 cents per m/Btu.[42] A still more recent contract confirms that capital costs of nuclear plants now are thoroughly competitive. Commonwealth Edison has contracted for a 700–800-mw unit at a per kw price well below that quoted for a 1,000-mw unit. It is contemplated that installed cost including land, interest during construction, and training will be about $110 per kw and that energy can be delivered at below 4

[40] Jersey Central Power and Light Co., *Oyster Creek Report, op. cit.* The ranges of capacity and cost reflect uncertainty concerning the steam output of the plant as well as differences in start-up cost, costs after the shakedown period, and costs in later life at lower plant factors. The plant will be designed to accommodate the higher output should the reactor prove capable of producing this amount of heat, while the lower figure represents the guaranteed capacity of the plant.

[41] However, it is worth noting that the British Central Electricity Generating Board has rejected an American bid for a nuclear plant at Dungeness B, citing the lesser availability of the American design because of its need to close down for refueling. The British design accepted is expected to produce power at 5.3 mills per kwh based on a 75 per cent load factor. An alternative computation of 30-year life and 85 per cent load factor brings this down to 4.4 mills (*Nucleonics*, September 1965 [New York], p. 25).

[42] The prices quoted include the following:

Size (mw)	Dollars per kw
300	156
500	131
600	122
1,000	107

(*Nuclear Industry*, October 1964 [New York], p. 8.)

mills per kwh.[43] Even if quoted prices are not duplicated on future bids, an added $20 or so per kw of capital costs with other factors unchanged would involve only an additional ½ mill in the cost of energy per kwh. Such a plant could be built or operated in any advanced country, and it is difficult to see why costs should differ markedly from those in the United States. Thus, for advanced countries, a sort of universal ceiling on energy costs in the neighborhood of 4–4.5 mills per kwh is a rather near-term prospect.[44] Over a longer period, still further economies may be anticipated. The technology is new and the normal expectation would be for declining costs as added experience is gained.

An obvious possibility for future economies is in nuclear fuel cost. Nuclear fuel, while comparatively inexpensive, is not free. The AEC study assumed fuel costs equal to 2.12 mills per kwh for 1966 operation, declining to 1.0 by 1980.[45] In addition, nuclear plants are expected to have somewhat higher nonfuel operating costs running 0.3 mills per kwh over coal plants in 1966 and 0.1 mill in 1980.[46] Power in this price range means that if such advanced countries, especially Europe and Japan, should decide to develop their domestic aluminum industries, they could do so without great cost penalty.

It is instructive to compare the costs for these recent contracts with those foreseen in other studies. The AEC study published in 1962 foresaw near-term capital costs of $160–$190 per kw for large nuclear stations—$35–$40 per kw more than for conventional plants.[47] Under those circumstances, they believed nuclear stations could generate at about 6 mills per kwh, competitive with fuel at 36 cents per m/Btu. In fact, as we have seen, the stations described above are considerably below this level of capital costs and are expected to provide energy about 2 mills cheaper per kwh. Moreover, the AEC calculations were based on an 80 per cent plant factor, but a plant serving the aluminum industry could expect to

[43] *Ibid.*, February 1965, p. 3.
[44] Philip Sporn, not known as a nuclear power enthusiast, concludes that an 800-mw nuclear plant can now be built for $123 per kw to yield power at 4.4 mills per kwh or competitive with coal at 24 cents per m/Btu (Philip Sporn, *op. cit.*, p. 24).
[45] U.S. Atomic Energy Commission, *op. cit.*, pp. 61-62 of Appendix. Such a fuel cost is equivalent to 12-25 cents per m/Btu.
[46] *Ibid.*, p. 59. These estimates imply costs well below 4 mills per kwh by 1980 if capital costs fall in the $120 range.
[47] *Ibid.*, p. 33.

do somewhat better. In general, the higher the plant factor and the larger the plant, the more competitive nuclear stations become. Large-scale aluminum smelters assure a high plant factor and can absorb the output of sizable nuclear generating stations. Where other markets for power are available, a smelter might absorb only a portion of the output but permit still larger and more economical generating stations to be built.

Another analysis by the National Power Survey published in 1963 shows that as of 1967 a 300-mw nuclear station could expect capital costs of $170–$190, equal to 0.7–1.1 mills per kwh more than from a coal station costing $130 per kw.[48]

In the somewhat more fanciful category lie the power costs projected by the American task force report on nuclear-powered sea water conversion plants.[49] In this report it was suggested that giant nuclear stations could be built which would take advantage of the fact that high-pressure steam is best used to generate electricity while spent low-pressure steam from the turbines is most suitable for sea water conversion. Such plants, with the energy component bearing a fixed charge at the commercial rate of 14 per cent, were deemed capable of providing power in the 3.6–4.3 mills per kwh range by 1975 for plants of 600–1,650 mw, while still larger plants might attain rates of 2.8–3.1 mills at sizes above 5,000 mw. Obviously such rates, if attained, would be attractive to the aluminum industry. Moreover, the absorption of this amount of power might present such difficulties that the attraction of power-using industries would be essential to the project. Few markets for converted sea water and electric power on this scale are apparent. The most obvious possibility is southern California, where markets for power, water, and metal are found together and access to Australian alumina could be had over a direct ocean route.

It must be borne in mind that nuclear technology is still in its infancy. The types of reactors which now are on the threshold of competitiveness with conventional steam plants are comparatively crude and their operating characteristics, while not well known, very probably are conservatively estimated. Despite this immature state of the art, the best projects already begin to compare favor-

[48] Federal Power Commission, *National Power Survey, Part I, op. cit.*, p. 88.
[49] Office of Science and Technology, *An Assessment of Large Nuclear Powered Sea Water Distillation Plants*, A Report of an Interagency Task Group (Washington: U.S. Government Printing Office, 1964).

ably with conventional plants in the United States where fuel costs traditionally are low. Similar technology transferred elsewhere could be still more attractive, while expected improvements in capital costs and fuel burn-up rates will cut nuclear costs. Therefore, the significance of nuclear power is that in the near future it provides an attractive limit to power costs in developed countries which now can be in the 4-mill range and which by 1980 should fall below this figure.[50] Since very little hydropower can be developed at costs of less than 2 mills per kwh, the potential advantage to be had through favorable hydropower sites in underdeveloped countries cannot be more than 2–3 mills in the very best of circumstances, and in most instances the advantage over thermal power in advanced countries will be less than 2 mills per kwh.

HYDROPOWER

The floor to the cost of thermal electricity is set by the cost of boilers and turbines and the final cost above that level is most heavily influenced by fuel costs. In the case of hydroelectric developments, under favorable circumstances the floor may be much lower since no boilers are needed and fuel costs are nil. In practice, however, extensive civil engineering works commonly are needed to store or divert water, often in sites where work is difficult, and it may also be necessary to transmit the power some distance to sites where smelters can be located most conveniently. Because of the wide variations encountered in stream flow, physical circumstances of the site, and convenience of location, it is almost impossible to generalize about hydroelectric power costs, but in nearly all circumstances the capital costs per kw are higher than for a steam plant of efficient size; this means that interest rates and period of amortization become especially critical variables.[51] At the same time, however, because of low maintenance costs and the

[50] See U.S. Atomic Energy Commission, *op. cit.*, Appendix, which anticipated a 6.2-mill figure for 1966, dropping to 3.8 mills by 1980. The pace has been faster than expected and the 1980 target figure thereby gains readier credence.

[51] E.g., for 17 new plants built in the United States and reported, capital cost per kw ranged from $198 to $464 (Federal Power Commission, *Hydroelectric Plant Construction Cost and Annual Production Expenses*, 6th and 7th Annual Supplements, 1962 and 1963 (Washington: U.S. Government Printing Office, 1965).

absence of fuel charges, operating costs are very low and the cost of energy may be lower than for conventional thermal power.[52]

The aluminum industry traditionally relied upon hydropower, with the earliest plants in the United States, France, and Switzerland being based on this source of energy. Only after World War II was extensive use made of thermal electricity in the United States and Europe, using gas in France and the U.S. Gulf Coast, coal in the Ohio Valley, and lignite in Germany. These shifts were indicated because of the shortage of well-located and inexpensive hydropower within the national economies of these countries. While these trends have drawn attention from hydropower, the aluminum industry still seeks the advantage of very inexpensive hydro wherever possible, even though the site may lack other raw materials or markets. Thus, the expansions in British Columbia and Norway have been based on the continued attraction of low-cost hydropower. Within developed countries most of the well-located and very low-cost hydropower already has been developed or, in any case, will find more attractive uses in other applications.

In less-developed countries this is not the case, however, and much attention has been given to the possibility of locating smelters in those areas where potential low-cost hydropower abounds. Often the initiative comes less from the industry itself than from an interest in economic development, whether originating with the national government, with former colonial governments that maintain a continuing sense of responsibility, or with purely international agencies. Such locations may be closer to sources of raw materials but often are far from markets, have high construction costs, lack skilled labor forces, and present an unstable climate for investment. So far there has been no stampede toward location in such countries. Nonetheless, as world demand for metal increases and remaining hydro sites in developed countries are exploited, the lure of inexpensive power in areas now passed over may grow. Therefore, it is necessary to consider where remaining hydropower is to be found and how attractive it may be to the industry from a cost standpoint, including cost of transmission where indicated. Such a discussion is necessarily woefully incomplete for a number of reasons, but it will help to give some dimension to the problem.

Hydropower potential depends upon the volume of water avail-

[52] At major TVA plants, for example, production expenses in 1963 typically were under ½ mill per kwh. (*Ibid.*, pp. 122-37.)

able and the distance which it falls. This relationship may be expressed by means of a formula which measures the gross theoretical potential for any hydraulic situation.[53] In no case could the gross theoretical potential be realized, because water wheels do not attain 100 per cent efficiency, nor can the full gross head be employed. More important are the losses attendant upon irregularity of stream flow and the mounting economic cost of utilizing higher percentages of the potential.

In principle, all water available on the surface (except evaporation and seepage) might be channeled to water wheels and converted to power. In practice it is necessary to identify potential power sites having appropriate physical conditions and stream flow such as to make development eventually practicable. Potential measured on this basis remains somewhat ambiguous, since no parameters of costs or demand prices for power are specified, but for our purposes figures assembled in this way provide at least some guide.

It is customary in measuring hydro potential to express it in terms of percentages of time which the potential will be available. Because of variability in stream flow, the potential available on an essentially firm basis (95 per cent of the time, or Q95) ordinarily is much smaller than that available only half the time. Since aluminum is a continuous-process industry with heavy costs attending temporary shutdowns, the 95 per cent basis is the one which should be considered here in appraising sites. Availability may be affected by modifications of the stream flow—the most obvious case being the use of storage to regulate flow. It is proper to use modified stream flow in determining availability, although this also involves ambiguity because of the lack of data on the cost of such modification.

Series compiled by the U.S. Geological Survey provide an overall indication of water power potentials for different areas.[54] The series are incomplete, however, since for many countries the necessary surveys simply have not been made. Moreover, in few cases

---

[53] 0.085 Qh where "h" is head in terms of feet and "Q" is flow in terms of cubic feet per second or 9.8 Qh where head is in meters and flow is in cubic meters per second (World Power Conference, op. cit., p. 39).

[54] U.S. Geological Survey, Summary of Developed and Potential Waterpower of the United States and Other Countries of the World, 1955-62, Circular 483, by Loyd L. Young (Washington: U.S. Government Printing Office, 1964).

have sufficiently detailed investigations been made to indicate potential costs of the power or the other uses for it. The figures are
of limited use also because gross potential is greatest where precipitation and terrain provide sizable flows and heads, while only
those areas not too distant from sea coasts or water transport provide the best circumstances for the aluminum industry.

Some of the world's greatest water-power potentials are found
in the Soviet Union, especially on the great Siberian river systems.
These are so remote from markets and materials that it is question-
able whether metal can be produced there for sale competitively
in non-Communist markets. Other large-scale resources are found
in the Congo, Canada, India, the United States, New Guinea,
Indonesia, Norway, Brazil, and the Malagasy Republic.

It must be borne in mind that a single economic-sized smelter
does not require truly massive blocs of power. A smelter of 100,000
tons does not suffer greatly from diseconomies of scale and can be
supported on about 200,000 kw of generating capacity. Thus,
countries with more modest power endowments than those cited
above, but which have low-cost, well-located hydro potential
accessible to the sea and for which other domestic demand is
slight, may prove to be more advantageous locations. Iceland,
Venezuela, New Zealand, Guinea, and others may fall into this
category.

No great amounts of unutilized, low-cost hydro are available in
Western Europe or the contiguous territory of the United States.
The only major exceptions are Norway (which offers power at
attractive cost) and Yugoslavia. However, Alaska and Canada both
have considerable potential. Very little of the hydro potential of
South America has been developed—the sole exception being
Brazil which nonetheless also has much which remains undeveloped. The great rivers of the Indian subcontinent have barely
been touched as a source of power. Much of this potential is poorly
located at remote and inaccessible sites (the Brahmaputra Bend,
for example) and higher uses will absorb the greater portion of
that more convenient power which will be developed soon. The
Mekong Basin could develop a great amount of hydropower but
most of it again is poorly located. Small amounts could be developed in New Zealand and the potential of New Guinea is considerable, though not well known. Africa and Malagasy are the
areas where really great unused potential is available. An enor-

mous potential is available in the Congo at a well-located site. Significant amounts are found in Malagasy, Cameroon, Gabon, and Congo (Brazzaville), while lesser but interesting sites are available in Guinea. Other sites are found well in the interior but, given the terrain and state of transportation in Africa, they need not be discussed here as prospects for the intermediate future.

It is possible to piece together estimates of hydro potential for many areas of the world based on U.S. Geological Survey reports or on individual country studies. It is quite another matter, however, to determine the significance of this for the aluminum industry because of the shortage of data on other prospective uses for the power and the costs of developing it. Therefore, although the following review of potential is useful in alerting us to various possibilities which should be considered, it will give us no very confident basis for predicting which ones may be selected.

HYDRO POTENTIALS BY AREA

In assessing hydro potential and some of the major projects which have been mentioned as possible sites for aluminum plants, it is useful to distinguish between potential for general power market supply and that more specifically available to the aluminum industry.[55] In developed countries where power demand is dense and distances to power markets reasonable, expansion of the aluminum industry cannot be said to depend upon specific power sites or projects. Here the important consideration is the overall power potential and expected growth in power demand. In less-developed countries, also, small locally oriented smelters may be built which are not tied to any particular hydro project. It is for the case of a large export smelter distant from other power markets that the specific project assumes importance.

North America

Firm hydro potential of North America is estimated at 90,065 mw gross concentrated in Canada with 43,240 mw, the United

[55] In figuring potential, the gross theoretical potential at Q95 as reported by Young will be employed except as otherwise noted (*Ibid.*).

States with 34,000 mw, and Mexico with 6,375 mw. Installed capacity exceeds the firm power total in the United States, is one-half in Canada, and one-third in Mexico (Table 21).[56]

Within the contiguous United States, the aluminum industry cannot hope to find isolated hydro sites which might be developed at low cost. This is due less to the exhaustion of hydro potential than to the growth and integration of the U.S. power market. More than two-thirds of the firm gross potential is found in the Mountain and Pacific states where demand currently is light. While additional

Table 21. Estimated Gross Theoretical Potential at Q95 and Installed Hydroelectric Capacity by Country and Continent

Area	Estimated Gross Theoretical Potential (mw)	Installed Capacity (mw)[a]
United States	34,000	38,600
Canada	43,240	20,315
Costa Rica	1,050	80
Greenland	750	10
Guatemala	1,575	30
Honduras	1,050	6
Mexico	6,375	1,900
Nicaragua	825	9
Panama and Canal Zone	525	65
North America Total	90,065	61,230
Argentina	4,050	400
Bolivia	2,700	110
Brazil	15,000	3,850
Guyana	2,700	6
Chile	9,500	688
Colombia	4,050	585
Dutch Guiana	825	U[b]
Ecuador	1,500	40
Paraguay	2,100	.4
Peru	4,800	450
Venezuela	3,225	500
South America Total	50,750	6,865

[56] Installed capacity may exceed firm power available in cases where hydro is used for peaking purposes. The figures on installed capacity are included here as a crude indication of the extent of development, but it must be recognized that the geographical disposition of potential may permit some to remain undeveloped even where installed capacity far exceeds potential.

Table 21 (Continued)

Area	Estimated Gross Theoretical Potential (mw)	Installed Capacity (mw)[a]
Angola	4,250	120
Federal Republic of Cameroon	4,800	159
Central African Republic	3,500	3.5
Congo	97,000	763
Dahomey	600	U
Egypt	375	365
Ethiopia	4,250	8.5
Gabon	6,000	18.6
Ghana	1,500	42
Guinea	500	20
Ivory Coast	500	20
Kenya	1,500	26
Liberia	4,250	3
Malagasy Republic	14,300	24
Mali	750	1
Mozambique	3,750	70
Niger	500	U
Nigeria	9,500	20
Southern Rhodesia, Zambia, and Malawi	4,680	810
Senegal	500(E)[c]	U
Sierra Leone	2,000	U
Equatorial Guinea	750	U
Sudan	750	U
Swaziland	700	U
Tanzania	3,000	40
Chad	1,000(E)	U
Uganda	3,000	121
Africa Total	176,677	3,185
Afghanistan	525	111
Burma	3,750	7.5
China	40,000	U
Formosa	750	560
India	31,000	2,450
Japan	9,000	14,000
Korea (North and South)	2,250	1,350
Netherlands New Guinea	7,875	4
Pakistan	6,560	346
Philippines	1,500	350
Thailand	3,000	30
Turkey	1,200	500
U.S.S.R.	48,000	[d]
Vietnam Republic	4,500	.15
Asia Total	160,826	19,992

Table 21 (Continued)

	Estimated Gross Theoretical Potential (mw)	Installed Capacity (mw)[a]
Australia	750	2,000
Indonesia	9,000	200
New Guinea	4,000	5.6
New Zealand	3,750	1,550
Papua	1,100	3
Oceania Total	18,600	3,860
Austria	3,000	3,600
Czechoslovakia	525	1,080
Finland	750	1,730
France	4,000	10,900
Germany (Federal Republic)	1,640	3,500
Iceland	525	110
Italy	4,500	12,700
Norway	7,500	7,600
Poland	750	260
Portugal	700	1,400
Rumania	1,500	230
Spain	2,625	4,850
Sweden	5,000	8,300
Switzerland	2,250	6,400
U.S.S.R.	15,000	19,000[e]
United Kingdom	562	1,350
Yugoslavia	2,400	1,600
Europe Total	54,687	85,806
World Total	551,605	180,938

[a] Capacity as of December 31, 1962.
[b] U=unknown.
[c] E=estimated.
[d] See U.S.S.R. in Europe.
[e] Includes Asian U.S.S.R.
Source: U.S. Geological Survey, *Summary of Developed and Potential Waterpower of the United States and Other Countries of the World 1955-62,* Circular 483, by Loyd L. Young (Washington: U.S. Government Printing Office, 1964).

power still can be developed in the Pacific Northwest, by now it is clear that this area will not retain a surplus of firm power into the remote future because, among other reasons, the prospects for major transmission connections with the rest of the country will end the previous isolation of this market.[57] These trends will affect the thinking of power marketing agencies and are likely to make sales to the aluminum industry appear less attractive than at present. Recent power contracts in the Northwest are advantageous to the aluminum industry because the price of Northwest power still remains low and because the companies expanding there (Harvey and Intalco [Péchiney]) can avoid the capital expenditures for power facilities.[58] However, the Northwest is distant from the major U.S. markets, and the interest of aluminum firms in locating there can be expected to dwindle rapidly if its power price rises in the future even to moderate levels.

The fact is that most postwar expansion of smelting in the United States has been based on thermal power sources. As has been noted, these resources are ample and are located more conveniently with respect to markets and raw materials. In combination with reduced power requirements per unit of output, these factors diminish the attractiveness of the Northwest.

In the eastern United States hydropower is provided to aluminum plants by TVA and the New York State Power Authority. However, there is no surplus of hydropower. In the case of TVA, additions to the supply must be figured at rates for thermal power and this is reflected in their price of over 4 mills.[59] The New York State Power Authority likewise charges a rate no better than that of a well-sited thermal plant.[60]

Considerable hydro potential is to be found in Alaska. Young lists a theoretical total at Q95 of 2,385 mw. Attention has focussed on the Rampart project, which proposes an ultimate development of 3,450–3,904 mw of firm power based on a storage dam on the Yukon which, under the most liberal financing assumptions, is expected to deliver power priced in the 3.3-mill range at tidewater

---

[57] Federal Power Commission, *National Power Survey, Part I, op. cit.,* pp. 256-65.

[58] Sales price of Bonneville power to aluminum plants in 1963 was 2.03 mills/kwh of firm power (U.S. Department of the Interior, *Annual Report, 1963,* p. 248).

[59] The minimum charge figures out to 4.12 mills per kwh in a recent contract with Consolidated Aluminum Corp. Actual prices are somewhat higher at Reynolds' plants and Alcoa (Tennessee Valley Authority *Annual Report, 1962,* pp. A34-A35, A43).

[60] Correspondence with New York State Power Authority described a minimum rate of 4.04 mills per kwh.

for prime load.[61] Realization of this project with its attendant cost figures is premised on the assurance of the location of aluminum or other electroprocessing industries in a remote environment and on a large scale. At least one major producer has expressed disinterest in such remote sites, and one variant of the project has included the possibility of long-range transmission to the Seattle area.[62] Despite this expected favorable power cost and legal and institutional climate, capacity in Alaska faces high construction and transportation costs which impair its attractiveness. An earlier study concedes high construction costs extending into the future.[63] In any case, the project is not expected to progress to the point where it will be a major supplier of aluminum before 1980. The Interior Department report shows 100,000 tons by that year.[64]

Canada still has vast supplies of unharnessed hydro potential in British Columbia, Yukon, Quebec, and Labrador which can be developed at low or moderate cost.[65] In eastern Canada the demand from the United States and Ontario is such that it appears likely to bid for power from the better located Quebec and Labrador sites. Discussions of the Churchill Falls project in Labrador, from which it is proposed to transmit power to New York City, is a dramatic illustration of this possibility. Much of the remaining hydro potential of eastern Canada is on streams flowing into Hudson and St. James bays, implying either a very severe climate for new plants or very long transmission of power. In view of the potential market for power in eastern Canada and the United States, it seems improbable that additional, truly low-priced energy will be available to the aluminum industry in this area.

In British Columbia the supply–demand relationship is different. A huge potential supply in a lightly populated area is far from other markets in either the United States or Canada.[66] The geology

[61] U.S. Department of the Interior, *Rampart Project, Alaska: Market for Power and Effect of Project on Natural Resources*, Field Report, Vols. 1-3 (Juneau, 1965), p. 495, also p. 818a. These terms exclude any charges for fish and wildlife mitigation, assume 3 per cent interest, deferred repayment of unused capacity of dam, and repayment of investment not to be completed until 50 years after the last investment is made.

[62] *Ibid.*, pp. 745-59.

[63] U.S. Senate Committee on Public Works, *op. cit.*, p. 37.

[64] U.S. Department of the Interior, *Rampart Project, Alaska, op. cit.*, p. 682.

[65] Canada, Department of Northern Affairs, *Water River Resources of Canada*, ‡2641 (1959).

[66] A Canadian industry source says 15,000 mw or more is available. Department of Northern Affairs, *ibid.*, estimates 18,200 hp = 13,576 kw, p. 2. HVDC

of the area somewhat complicates development, but apparently costs will be in the moderately low range.[67] The major development at Kitimat has illustrated the problems of development in British Columbia where "beachhead costs" of developing port and townsite, and the problem of attracting labor in combination with the difficult geological conditions all have made for an expensive undertaking. There has been no rush on the part of aluminum firms to secure other sites in the area. Despite the problems, however, the moderate cost of power, hospitable investment climate, and access to the sea favor the area for eventual further development.

The only other sizable hydro resources in North America are found in southern Mexico. There has been no discussion of the use of this power in the aluminum industry and it seems more likely that it will be put to general commercial and industrial use in a rapidly growing economy rather than be used to support an export smelter.

South America

Total resources in South America at Q95 are figured at 50,750 mw and present installed capacity is only 6,865 mw. In truth, potential and costs in South America are not well known. The greatest potential is found along the Andean spine and therefore is distributed among all of the countries of this area. However, Brazil and Venezuela have major potential resources in other locations.[68]

Under present conditions, sites on the eastern slope of the Andes in the northern part of the continent seem too remote for consideration. It is believed that the power available on both slopes of the Andes in Patagonia and Tierra del Fuego would be relatively inexpensive, but again the areas, while accessible to the sea are remote from materials and markets, and it would require substan-

could, however, deliver power from British Columbia to California in the next 10 years if nuclear power cost does not drop to a level precluding this. M. W. Stoddart, *The Role of HVDC in the Electrical Development of Canada*, pp. 5-6 (mimeo.).

[67] U.S. Senate Committee on Public Works, *op. cit.*, p. 67, suggests 3-5 mills. A Canadian industry source puts cost in the Kitimat range and competitive with Rampart.

[68] See United Nations, *Hydroelectric Resources in Latin America: Their Measurement and Utilization*, ST/ECLA/Conf.7/L.3.0, January 12, 1961, Map IIb.

tial infrastructure investment to locate a plant there.[69] Moreover, seismological conditions in Chile in particular might impose an additional constraint.

Brazil has an aluminum industry and a growing market. It also has power potential to permit expansion of its industry, but much of it is poorly located.[70] Moreover, demand for power for general use will require much of the better-located potential. The more probable evolution in Brazil is that it will continue to expand output to meet local needs.

Remaining attention in South America focusses on four countries. Surinam offers some promise. Its resources are listed at 825 mw by Young. Private studies by German interests suggest that substantially more than this is available at low cost, but this evaluation is challenged by others. Given the excellent juxtaposition of bauxite and power in combination with access to the EEC market and a record of governmental stability, the ingredients seem to be present for at least some expansion beyond the moderate-sized smelter now being constructed. Guyana has more potential and, in addition, has large supplies of bauxite. The cost of developing the power is unknown, but so far there has been little speculation about a smelting industry in the country—perhaps because Guyana has communal strife and a political climate unlikely to be conducive to investment for some time.

In Peru, the Mantaro project originally was discussed in terms of possible location of an aluminum smelter. Subsequent discussion suggests that it will be so phased over time that the power can be absorbed into general use in urban areas of western Peru. There has been no recent mention of use of this project for aluminum reduction.

The Caroní in Venezuela offers one of the more promising possibilities in South America. It is hoped to have 1,750 mw installed by the early 1970's and ultimate development of this site is estimated on the order of 6,000 mw. With ready access to Caribbean and South American bauxite and with a deepwater channel available via the Orinoco, this low-cost site offers many advantages.[71] The country has attained a significant rate of economic growth.

[69] U.S. Senate Committee on Public Works, op. cit., p. 68. Chile has a potential of 9,500 mw, most of it in the southern part of the country.

[70] Young lists 15,000 mw available at Q95 and installed capacity of only 3,850 mw (U.S. Geological Survey, Circular 483, op. cit.).

[71] While the price of power to the aluminum industry is still unannounced, it is expected to be on the order of 3 mills per kwh.

Aluminum production would be part of an overall regional development program and therefore would not bear the major costs of harbor, townsite, and other infrastructure. Labor is expensive in Venezuela, but for a plant of efficient design this is not a serious drawback. Moreover, should the Latin American market grow rapidly within a LAFTA framework, Venezuela would appear best positioned by reason of resources and location to become the major Latin American supplier. Initial plans are for only a small plant, but long-term plans are very ambitious. Prospects for the Latin American market are strong enough that there is some prospect for sizable development in Venezuela.

Africa

Africa has a listed hydro potential higher than any other continent, with 176,677 mw at Q95 of which a mere 3,185 mw is developed. With its considerable bauxite resources as well, it has been the object of much speculation by the aluminum industry but so far it has had comparatively little smelter investment.

By far the major power potential, well over half of the total, is found in the Congo, which also has the world's most remarkable power site. The next greatest concentration is in the Malagasy Republic. Other major water powers are in Nigeria, Gabon, Cameroon, Angola, Southern Rhodesia, Ethiopia, and yet other countries. Generally speaking, these constitute the tropical belt of the continent, there being very little hydro in the Maghreb countries (Algeria, Tunisia, and Morocco) or in South and Southwest Africa.

With potential on the scale suggested here, there inevitably are many possible sites capable of supporting smelters of economical size. Moreover, if allowance is made for transmission of any distance, the number is quickly magnified. Attention has focussed on the one existing smelter in Cameroon and the one being constructed in Ghana. Beyond this, serious mention has been made of prospects in Angola, Guinea, and the Congo (Brazzaville), as well as the possibility of a local smelter in Egypt.

Both the existing smelter in Cameroon and the planned smelter in Ghana draw power from projects which were initiated with other aims in mind. The Cameroon project was a colonial development scheme which then offered a major part of the power to an aluminum company at a favorable price.[72] In Ghana, a substantial portion of the power revenues will be derived from sale of power to users other than the industry.[73] In both of these cases the existing or proposed plants utilize substantially the supply that will be available for aluminum production at these sites, so no major expansions are envisaged.

Egypt has talked with various foreign interests concerning the possibility of a smelter to use part of the power from the Aswan Dam. With a push from the state this may be brought about as a local or regional plant supplying the North African area. Again the political climate does not appear favorable for large private investment, and Egypt is unlikely to become a major export supplier.

A project on the Cuanza in Angola has excited the interests of foreign firms and at one time seemed likely to be built. However, slack demand postponed it and the interest of the foreign firms has shifted elsewhere as political turbulence makes the area less promising for private investment.

The French government has conducted investigations of power potential on the Kouilou in the Congo (Brazzaville) and concluded that the Sounda Gorge could be developed to deliver 800 mw at Pointe Noire on the coast—a distance of 93 km. Computed on the basis of International Bank for Reconstruction and Development (IBRD) financing terms allowing for a fixed charge of 8.5 per cent, the French concluded that power could be delivered at approximately 2 mills per kwh.[74] On this basis a project for a 250,000-ton smelter has been advanced, but again no action has been taken despite the interest expressed as long ago as 1958. The present political situation in the area is not likely to encourage realization of this project.

On the Konkouré in Guinea another site has been investigated by the French government. This site is 130 km from the port of

---

[72] World Power Conference Sectional Meeting, *Utilization de l'énergie disponible d'une chute d'eau pour le traitement de l'aluminium en Afrique centrale* by M. de Verteuil and J. Ribadeau-Dumas (Belgrad, 1957).

[73] Based on discussion with officials of the International Bank for Reconstruction and Development. During various stages the smelter takes about 60 per cent of the power but will yield only 30 per cent of the anticipated revenues. Power is priced to the smelter at 2.63 mills. Charges to other industrial uses may be as high as 15 mills.

[74] Based on data furnished by Overseas Service of Électricité de France.

Conakry and near the Fria alumina plant. It is proposed to develop 360 mw at the site. Under the same assumptions of IBRD financing it is estimated this could be done for about 3 mills or a little more.[75] Given the most unusual juxtaposition of bauxite, a large, going alumina plant, and power at this price, it might be supposed that the project would be eagerly embraced by the aluminum companies. Instead, it has languished and there are no current plans to go forward. So far as is known, the intrepid Harvey venture into the Boké bauxite mining project in Guinea, with its implied commitment to an alumina plant, does not extend as far as the building of a smelter there.

The French have made somewhat less detailed investigations of the hydro resources of the Malagasy Republic. Reportedly the costs in this case would be in the same range as at Kouilou and the potential supply much larger. Since the island is not very wide, distances to the sea should be well within acceptable transmission range. The political stability of the area since independence has been encouraging, yet no mention has been made of aluminum projects for the island. It is remote, both from markets and known ore sources, but it may receive attention later if some of the better known sites must be passed over.

Eventually anyone casting his eyes upon the power potential of Africa must come to rest on the Congo and its fabulous Inga project. In this instance a unique combination of a huge river of stable regimen, a sharp drop in elevation over a short distance, and close proximity to deep water combine to make a power site unique in the world. Unhappily, the very magnitude of the project so far has done much to defeat it, and the distressing political situation in the area seems likely to frustrate it for some time into the future. However, it is of such scale that if it is ever to go forward it must be on the basis of supplying aluminum to the world. Once started it would exert a powerful attraction to smelters if proper investment security could be assured.

The site at Inga is 40 km from deepwater at Matadi on an elbow of the Congo where the river drops 96 meters over a series of rapids. Because of its immense drainage basin encompassing a variety of climates, the Congo enjoys a very stable flow and can be used without regulation. At ultimate development the project would divert the entire low flow of the river, cutting across the

[75] *Ibid.*

bend by way of natural channels in part, to develop power of 25,000–30,000 mw. Compared with our estimates of total new power needed by the industry by 1980, the scale of this project can be appreciated. No other important uses for power exist in the area or within the range to which it might be transmitted.

Numerous studies of the project have been made and quite varied plans and costs have been suggested. One of the most difficult problems is to design an economical first-stage development which will not compromise complete development. Apparently this can be done, but only at costs somewhat above those for the nearby Kouilou project which is of more manageable scale.[76] As the size of the project increases, it seems likely that costs will drop under 2 mills per kwh based upon a fixed charge of 8 per cent.[77]

Various estimates have placed the cost at full development between 1 and 1.6 mills per kwh, all based on studies several years old with somewhat differing but always favorable assumptions concerning financing. While this is certainly a very attractive figure, it must be borne in mind that the investment in power facilities alone to develop 25,000–30,000 mw would be $3–$4 billion.

The development of even a fraction of this would be sufficient to supply a major portion of the expected increase in world demand for aluminum, but it would require investments beyond the capacity of any single firm. Therefore, a consortium of major firms would have to be envisaged, with financing of power facilities very probably done through an international agency. Moreover, since investment in smelter facilities will be substantially higher per ton of capacity than that in power, some form of guarantee of this investment also would seem to be necessary. There is neither urgent need nor great possibility of working out this form of arrangement in the near future, and major development of the Inga project seems destined to lie fallow for yet some time. However, the shadow of this very low-cost project inevitably must dim the prospects for numerous other conceivable proposals in the less-developed countries.

[76] E.g., at 800 mw, equivalent in size to Kouilou, the cost is figured at about 3 mills (Royaume de Belgique Ministere des Colonies, *Amenagement hydro-électrique du site d'Inga*, Rapport du comité des experts (1957).
[77] *Ibid.*

Asia

Gross theoretical hydro potential of Asia is estimated by Young at 160,826 mw at Q95, second only to Africa. About 90,000 mw of this is in Siberia and China, leaving approximately 70,000 mw in non-Communist areas of Asia. India has the greatest share of the remaining total with 31,000 mw, while major powers are found in Japan (mostly developed), Pakistan, the former French Indo-China areas, Thailand, and Burma.

Despite its large hydro potential, India is unlikely to offer attractive sites for export smelters. Much of the power is located far from the sea and indeed from population centers.[78] Moreover, the nation is hard put to find investible funds to build power for higher uses. Hydro will continue to be developed and applied to smelters producing for the local market, but this will be of scant world significance.

Much the same can be said of Pakistan, with the additional limitation that it is poorer in hydropower potential than India and so far has no smelter capacity. For reasons of status or foreign exchange it may acquire some local smelter capacity but can be excluded as a site for export smelters.

Japan has developed its low-cost hydro and is compelled to rely upon thermal energy for much of its additional power. It seems more apt to be an importer than an exporter of metal in the future.

Another expression of interest has occurred in Turkey where considerable hydro potential is found in the eastern part of the country. At present there appears some likelihood that a project may go forward since power could be delivered to more accessible areas at interesting prices.

So far as is known there has been no discussion of reduction plants in Burma or Thailand, despite their considerable hydro potentials. Rivers throughout Southeast Asia and Indonesia tend to be highly variable because of the monsoon season and, unless fed by remote snow fields, often require regulation.

The four lower riparians in the Mekong Valley with United Nations assistance have spent considerable effort in examining the potential of this basin and hope to develop the river on a multipurpose

[78] The greatest site is at the Brahmaputra Bend, quite inaccessible for an aluminum smelter.

basis. If they should succeed in surmounting the numerous obstacles, they would be able to provide large blocs of power perhaps usable in the aluminum industry. Unfortunately, most of this is well upstream on a river that is not continuously navigable. It would be possible to transmit power from the most accessible sites at Sambor or Stung Treng to Saigon or to downriver sites accessible to ocean shipping, provided civil peace in the area were established.[79] It is doubtful that any private firm would be interested in a major investment on this basis, but there apparently is hope that Japanese interests with government backing could be interested. This would have the advantage of a market connection in a country ill-equipped to meet its own needs. However, if purely commercial considerations prevail, due account taken of conditions in the area, the project seems very improbable, and the more so if nuclear energy becomes available to Japan in the 4–5-mill range.

Oceania

Turning to the remaining islands of Southeast Asia and to Australia, the greatest hydro potential is found in New Guinea, with the island total almost 13,000 mw (including 7,875 mw which Young puts in the Asian figures). Indonesia is credited with 9,000 mw and New Zealand 3,750 mw. Australia has comparatively little hydro, except in Tasmania, and will have none to spare for aluminum smelting.

The potential of New Guinea is very considerable and not fully explored. At one time there was interest in building a smelter in the Australian portion of New Guinea but the project currently is moribund. The availability of vast supplies of bauxite in northern Australia may ultimately revive this interest, but prospective investors would surely consider carefully an investment in one of the world's most primitive areas whose political status is apt to be questioned at any time.

The other interest in Oceania is in New Zealand where the government is proceeding with the development of the resources of South Island. Part of this power will be transmitted to the north, but negotiations for an export smelter in New Zealand, temporarily

[79] Studies are not complete but some Japanese sources estimated costs in the 2–3-mill range.

shelved when Australian capacity was expanded, now seem likely to be revived. Power costs will not be as low as some suggested for less-developed countries elsewhere in the world, yet the proximity to the sea and location in a stable political environment with good access to Australian alumina nonetheless make it of interest.

Indonesia has planned a small local smelter and has sought technical assistance and financing from Communist-bloc countries. Such a project, if built, will have no world significance.

Europe

Total gross potential at Q95 for Europe is listed at 54,687 mw. Of this amount over 18,000 mw is in Communist-bloc countries, leaving about 36,000 mw in the West. The greatest potentials of the West are found in Norway with 7,500 mw, Sweden with 5,000 mw, followed by Italy, France, and Austria.

European hydro potential is more fully exploited than in any other continent. It is typically the case that installed capacity exceeds firm potential and much European hydro is developed for peaking or seasonal use.[80] Only three countries have significant undeveloped water power in the cost range that would be of interest to the aluminum industry; these are Norway, Yugoslavia, and Iceland.

Although Young lists the firm potential of Norway at 7,500 mw, the Norwegians have more ambitious plans to develop their water powers through regulation and diversion. They have made a complete inventory of their potential and have classified it by categories of cost.[81] On this basis they figure they can develop about 130,000 gigowatt hours (gwh) or the equivalent of about 15,000 mw of developed power. More than half of this is expected to be in the lowest price class of under 3.1 cents per kwh of annual capital cost and as of mid-1964 there remained about 40,000 gwh or nearly 4,600 mw in this class still undeveloped.[82] Norway seeks to add

[80] Installed capacity outside the Communist-bloc countries is 64,700 mw.
[81] See, e.g., Sixth World Power Conference, *Developments in the Exploitation and Use of Energy in Norway*, Vol. I, Paper No. 1.2/23 (Melbourne, 1962).
[82] Discussion with G. Roald and F. Vogt of the Norwegian Watercourse and Electricity Board.

about 600 mw per year of new capacity and at this rate should succeed in harnessing its most economical potential by 1975. Under the IBRD terms available to Norway, they estimate that at 3.1 cents capital cost per annual kwh, the least expensive Norwegian power will be produced for just under 3 mills.[83] This compares with the present state price of 2.5 mills at the generating station.

Iceland is credited with a firm potential of 525 mw of firm power by Young. However, the technically exploitable potential is far higher, being currently estimated at 35,000 gwh per year, or about 4,000 mw, of which at least some share can be produced in the 2–4-mill range.[84] Seismological and ice conditions pose a problem, but it still seems likely that many hundreds of mw of capacity can be built in the indicated price range.[85] It is known that negotiations concerning the location of a smelter in Iceland have taken place and, according to a recent announcement, a commitment has been made. In this case potential power costs, proximity to markets, and investment climate combine to make the project worthy of consideration.

Yugoslavia is credited with 2,400 mw of firm power potential by Young. An outside study of certain hydro projects was made by the United Nations Economic Commission for Europe (UNECE) in 1953, in connection with proposals for seasonal exports of power to Western Europe.[86] Although the study focussed on power sites of interest for this seasonal export purpose, the estimated capital costs of between 4–5 cents per annual kwh for some of the projects suggest energy costs in the neighborhood of 4–5 mills.[87] Quite likely lower costs could be had for base load operation. Thus, Yugoslavia is one of the few European countries with undeveloped power in a cost range likely to be interesting. Of course, investment in this case would be via state entities and the metal, if intended for export, would face all of the problems of nonintegrated producers in attempting to penetrate foreign markets, plus the added prob-

[83] *Ibid.*
[84] Correspondence with Jakob Gislason, State Electricity Board. See also World Power Conference Sectional Meeting, *Economic Aspects of Utilizing Surplus Energy Resources in Iceland,* by Glúmer Björnson (Belgrad, 1957).
[85] Gislason in letter cited above suggests 20,000 gwh are economical but concedes the vagueness of the concept.
[86] See United Nations, *Prospects of Exporting Electric Power from Yugoslavia—Summary* E/ECE/192, E/ECE/EP 154 (Geneva, 1955).
[87] *Ibid.,* p. 6.

lem of skepticism concerning the stability of the supply from this source.

Up to the present, aluminum smelters typically have been located very close to the power so that little or no transmission has been required. Both for thermal and hydroelectricity this has meant that the primary energy source needed to be well located with respect to transportation in order to be attractive to the industry. There have been exceptions where less well-located sites have been used—notably some of the plants in the Soviet Union and Anaconda's plant in Montana—but in no case has long-range transmission of power been employed. As a consequence, power sites deep in the interior, unless served by a good inland waterway system, ordinarily have not been considered as a basis for aluminum production.

Yet it is apparent that if long-distance transmission is feasible for the industry, the number of potential power sources quickly multiplies. Nature infrequently is kind enough to provide waterfalls close to the sea—more commonly the larger heads are located well to the interior where the great continental mountain ranges concentrate precipitation while at the same time the elevation implies greater heads. Inland plant sites that are accessible to barge traffic, as is true of some United States and European plants, need not suffer so great a disadvantage, but in the absence of good transportation such sites become costly, if not entirely unfeasible. Inexpensive transmission of power from energy source to a more convenient production site therefore can become of great importance to the industry as the more accessible sources of cheap power are pre-empted.

Advances in transmission have another kind of significance to the aluminum industry. Heretofore, the industry has tended to locate in areas where power was cheap. Often the power was cheap not merely because of low production costs but also because of the absence of other uses for it. This has implied relatively isolated sites in some cases. However, long distance transmission provides the possibility of reaching an alternative market for the power and thereby bidding up its price. Thus, as the industry reaches out to new sources of inexpensive power, it may find them snatched away through advances in transmission which now place such power within the ambit of a larger market (Labrador and Norway, for examples).

In recent years there have been great advances in long distance extra high voltage (EHV) transmission. In the United States the common transmission voltages historically have been 230 kilovolts (kv) or less. Only around 1950 was 345 kv introduced and higher voltages are now planned or in operation in the United States and elsewhere. Ambitious Soviet plans for transmission of Siberian power envision long distance transmission at 500 kv or more. In the United States, the California–Pacific Northwest intertie will represent a new step involving both AC and DC lines of 500–700 kv. Meanwhile Quebec hydro is constructing transmission facilities to operate at 735 kv. There has been discussion of sending Labrador power into the United States by EHV.[88]

Power transmitted long distances must be carried at high voltage in order to minimize line loss, while the cost of constructing long-distance lines normally requires that they be used to transmit large blocs of power. At a given capacity, increased voltage increases the distance possible far more than proportionately, the power transmitted by a single line being proportional to the square of the voltage.[89] For a given distance, increasing the voltage also increases the capacity more than proportionately.[90] The prevailing rule of thumb for AC transmission is that power is most economically transmitted about 1 mile per kv.

Capital costs are the principal element in transmission. Costs of higher voltage lines do not increase in proportion to their added capacity. Thus, a 345-kv line can carry 9 times the power of a 115-kv line but will cost only 4–5 times as much.[91] The savings

[88] See, e.g., Daniel T. Braymer, speech to New York Society of Security Analysts, February 6, 1963, pp. 1-3; D. M. Farnham, *Apparent Cost Trends in Hydroelectric Developments and the Transmission of Electric Power,* paper presented to Canadian Nuclear Association Conference (Montreal, 1963), pp. 4-5. As late as 1962 the longest transmission line in the United States was the 287-kv, 265-mile line from Hoover Dam to Los Angeles (Federal Power Commission, *National Power Survey, Part I, op. cit.,* pp. 149-51).

[89] Braymer speech, *op. cit.,* p. 5.

[90] U.S. Senate Committee on Interior and Insular Affairs, *Report of the National Fuels and Energy Study Group,* Senate Document 159, 87th Cong., 2nd sess., September 21, 1962, p. 200.

[91] Braymer speech, *op. cit.,* p. 5. See also Federal Power Commission, *National Power Survey, Part I, op. cit.,* Chapter 10.

come from better use of right-of-way and conductors. Terminal equipment assumes a special significance where cost per unit of distance is in question because, while line costs for a given capacity line tend to be proportional to distance, terminal costs are fixed and therefore on a per mile basis diminish with distance.

The great importance of capital costs also emphasizes the significance of load factor in the cost of transmission. Since long distance EHV is suited to circumstances where load factors are high, it is therefore of special interest to aluminum.

Standard EHV long distance transmission voltages contemplated for the United States will be 345 kv, 500 kv, and 700 kv (nominal ratings).[92] Representative loadings for single circuits at these voltages are given by Stoddart at 390 mw, 890 mw, and 1,800 mw respectively.[93] The amount of power actually delivered diminishes with distance because of line losses. However, at appropriate voltages for the distances concerned these losses remain within the range of 5–10 per cent.

In the absence of any real experience, various efforts have been made to estimate the cost of long distance EHV, both AC and DC systems.[94] Stoddart worked out the capital cost of delivering of a given load at various distances by AC and DC lines of comparable capacity. In a very interesting chart he shows the distance which power can be delivered from generating site at 4 mills per kw under varying capital costs of generation, fixed charge rates, and types of transmission.[95] The results are not immediately encouraging. At 12 per cent fixed charge and $200 per kw capital cost of generation at hydro plants no power could be delivered as far as 200 miles for 4 mills. If fixed charges are considerably lower (implying government financing), then generating sites in the $200–$300 per kw class can deliver 4-mill power at great distances.

A related graph prepared by Farnham on the basis of a 10 per cent annual charge shows that at 90 per cent load factor power could be transmitted distances of 500–700 miles for little more than 1 mill per kwh—a figure that could prove interesting to the

[92] Federal Power Commission, *National Power Survey, Part II, op. cit.,* Advisory Committee Report No. 9, p. 83.

[93] *Ibid.,* p. 2.

[94] Among them are the Federal Power Commission, *National Power Survey, Part II, op. cit.,* Advisory Committee Report No. 10, p. 87, and Stoddart, *op. cit.*

[95] Stoddart, *op. cit.,* Chart 4.

aluminum industry given exceptionally favorable costs of generation.[96]

Much attention has been given to comparison of AC and DC transmission. The attraction of DC is that it does not face the same stability problems as AC at very high voltages and therefore can be used for longer-distance lines where AC has theoretical limits which will restrict it technically to around 1,100 miles. Although the technology of high voltage direct current (HVDC) is not well developed, it seems likely to prove more attractive for large blocs of power moved great distances without intervening service.[97] Thus, it may be of interest to the aluminum industry.

Before leaping to any conclusions concerning the application of EHV to the aluminum industry, it is well to keep several cautions in mind. First is the state of flux in EHV technology. There is very little experience with the higher voltages and attendant costs in AC, while still less is known about HVDC where costs of equipment never yet built can only be approximated. Developments could exceed expectations or they may prove disappointing. Second, it must be remembered that broad generalizations say nothing of the specific circumstances encountered. Terrain and weather greatly affect the design required to ensure high reliability of service, and costs of maintenance also vary with the circumstances. Some of the most attractive power sites present the most difficult conditions of terrain and arctic or tropical environment. If the power is for aluminum, these problems must be solved—usually at some cost—since high reliability of service is required. Finally, the attractive transmission costs are to be had only with very large blocs of power. Even the largest aluminum plants would not require all of the power deliverable by multicircuit (essential for reliability) lines at EHV. Thus, other markets for power ordinarily would need to be available within the transmission range.

Bearing these thoughts in mind, it would appear that there is likely to be little immediate application of EHV to the industry. No surpluses of inexpensive power potential exist in interior portions of the United States or continental Europe. Possibilities in Africa suffer from the absence of necessary supporting demand for power from other sources, as well as from the reluctance to make large-scale investments there. The chief near-term possibility

[96] Figure 4 in Farnham, *op. cit.*
[97] See Stoddart, *op. cit.*, Figures 1 and 2.

would appear to be the movement of Siberian hydropower to more convenient centers. Over a longer period, assuming the development of a favorable investment climate in less-developed countries accompanied by the exhaustion of moderate-priced power resources of developed countries, EHV could result in the development of inland sites in South America, Africa, and Asia, provided generating costs on the order of 2 mills are possible. If we figure that alternative conventional thermal or nuclear power will be available in developed countries in the 4-mill range or less by 1980, then even the best power sites in less-developed countries would have difficulty in delivering power at an attractive price if the power had to bear a transmission charge of over 1 mill. Nonetheless, if such countries present a stable environment for investment in the future and also have a favorable juxtaposition of raw materials, the advances in transmission technology will broaden the alternatives open to the industry.

CHAPTER TEN

RÉSUMÉ ON LOCATION

At this point the various factors affecting location may be pulled together. This will not be done in a detailed fashion. While it might be possible to estimate costs at various prospective sites on a current basis, for future conditions there is not sufficient information or certainty to justify detailed estimates. Moreover, for our purposes it appears unnecessary to study specific locations. Such an exercise would only obscure the major factors likely to shape location decisions during the period of concern. Also it would not be helpful in other situations where, in addition to the differences in cost, location will be influenced by considerations of industry structure and of company and national strategy whose force must be evaluated even though they cannot easily be reduced to monetary terms.

REGIONAL CAPACITY AND EXPECTED CONSUMPTION

As a preliminary step it can be assumed that existing plants will remain in production and that announced projects will be carried out. The latter will not prove to be precisely true, for some projects will be abandoned, but nonetheless this will give a rough picture of the industry as it is or will be over the next few years. In combination with the estimates of demand, it indicates the regional gap between foreseeable capacity and consumption which will need to be met by expanded local production or by imports.

The total existing and announced capacity in the non-Communist world as of early 1966 is shown in Table 22.[1] The striking aspect of

[1] Capacity figures are a useful guide but should not be taken too literally. Some plants regularly produce more than rated capacity, whereas others are

Table 22. Primary Aluminum Smelting Capacity, Existing and Planned, by Country
(thousand metric tons)

Country	Existing Capacity	Announced Plans	Indefinite Plans
United States	2,661	406	159
Canada	865	73	245
Mexico	20	—	50
North America Total	3,546	479	454
Austria	79	4	25
France	372	—	—
Germany	234	30	30
Italy	133	100	—
Netherlands	—	60	—
Norway	345	251	300
Spain	56	53	—
Portugal	—	25	—
Sweden	30	18	20
Switzerland	70	—	10
United Kingdom	35	—	—
Yugoslavia	51	—	150
Greece	21	57	15
Iceland	—	60	—
Europe Total	1,426	658	550
India	100	45	330
Japan	340	110	120
Kuwait	—	—	50
Indonesia	—	—	20
Taiwan	20	—	20
Turkey	—	—	30
Asia Total	460	155	570
Cameroon	53	—	50
Ghana	—	104	50
Angola	—	—	25
Congo (Brazzaville)	—	—	250
Guinea	—	—	150
Egypt	—	40	—
South Africa	—	—	20
Congo (Kinshasa)	—	—	500
Algeria	—	—	20
Africa Total	53	144	1,065

subject to power shortages and rarely equal rated capacity. Often improvements resulting in added output potential are slow to be incorporated in published figures. Finally, since the series was compiled from diverse sources, it was sometimes unclear whether the reported figures were in metric or short tons and some error on this account is likely.

Table 22 (Continued)

Country	Existing Capacity	Announced Plans	Indefinite Plans
Australia	96	48	18
New Zealand	—	—	120
Oceania Total	96	48	138
Brazil	35	33	80
Surinam	53	—	—
Argentina	—	—	20
Venezuela	—	10	500
Curaçao	—	—	30
South America Total	88	43	630
NON-COMMUNIST TOTAL	5,669	1,527	3,407
Communist Bloc[a]			
China	150	—	—
Czechoslovakia	54	—	—
East Germany	67	—	—
Hungary	65	—	—
North Korea	35	—	—
Poland	61	—	—
Rumania	23	—	—
U.S.S.R.	1,359	—	—
COMMUNIST-BLOC TOTAL	1,814	—	—

[a] No figures are shown for future expansion in the Communist bloc. However, Soviet plans apparently call for 1,900,000–2,000,000 tons of capacity by the late 1960's. Rumania may add another 100,000 tons.

Sources: Based on U.S. Bureau of Mines, *Minerals Yearbook, 1963; Metal Bulletin*, "Aluminium World Survey" Special Issue, December 1963; Curt Freiherr von Salmuth, *Handbuch der Aluminium Wirtschaft*, 1963, and subsequent press reports as of early 1966.

this table is the heavy concentration of existing capacity in North America (63 per cent), but the much faster relative growth to be expected in Europe on the basis of announced plans. A second feature is the scarcity of existing capacity in the underdeveloped world, although in terms of announced projects this area does somewhat better. On the basis of present plans, the most noticeable shift occurring is the relative growth of Europe, with over 43 per cent of announced projects representing about a 46 per cent growth in European capacity. Although heavily weighted by Norwegian projects, this activity in Europe must come as a shock to those who long ago had written Europe off as a site of future expansion. It is

227

not until we look at the vaguer plans for the future that the under-developed world assumes major significance; even this is somewhat deceptive because possibilities listed in this category may be super-seded by projects in developed countries based largely on thermal power for which no great advance consideration is necessary.

Taken together, the existing capacity and announced expansions total about 7.2 million metric tons outside the Communist bloc. Recalling the anticipated demand for new primary metal by 1980 of 12.9 million metric tons and allowing for excess capacity of 15 per cent, we arrive at some 8.0 million metric tons of additional capacity likely to be built between now and 1980 beyond that already built or announced.[2] This means that the capacity of the industry must be more than doubled by the target date of 1980.

A comparison of existing and announced capacity with antici-pated demand by region in the non-Communist world is given below:

	Existing and Announced Capacity (thousand metric tons)	Expected Demand for Primary, 1980[a] (thousand metric tons)
Europe	2,084	4,035
Latin America	151	525
United States and Canada	4,005	6,370
Asia	615	1,485
(Japan)	(450)	(985)
Africa	197	195
Oceania	144	290
Total	7,196	12,900

[a] Expected consumption figures from page 73 reduced to allow for sec-ondary metal amounting to 4.6 million metric tons. Secondary consumption was assumed to be distributed approximately in proportion to its present share of output in each country.

The table clearly shows that the gap between expected demand for primary metal and existing and announced capacity is greatest for Europe, with other major needs likely to occur in North America

[2] I.e., capacity of 15.2 million metric tons is needed to permit 15 per cent of capacity to be either idle or producing for inventory when consumption is at the 12.9-million metric ton level. Over the past 10 years consumption has rarely exceeded 90 per cent of year-end capacity and often has fallen below 80 per cent. Hence a target of 85 per cent does not appear unreasonable.

and Asia. Comparatively little new capacity will be required to meet the local needs of Africa and South America. Of course the table should not be read as showing opportunities for new smelters in each region; trade will continue to move metal between advantageous production centers and markets. All the table does is to juxtapose markets and capacity, thereby suggesting the dimensions of the problem of supplying the various regions.

Given the expected growth of the major markets beyond existing capacity in the industrial centers of North America, Europe, and Japan, and the location of bauxite in the Caribbean, Africa, and Australia, the problem of location of new smelters is sharply posed. Will Africa, South America, and perhaps Southeast Asia become major suppliers of metal to the industrialized areas to the north? As we have seen, they have not yet become such suppliers. Despite the fact that new projects are under way in Ghana and Surinam for export purposes, a far greater share of the expanding needs of industrialized countries is being met by expansion in their own region. Expansions in Japan, the United States, Canada, Norway, and Germany all have followed the initiation of the projects in Ghana and Surinam and, while still more are in prospect for the industrialized areas, the projects for export smelters in the less-developed countries have had no recent imitators.

SIGNIFICANCE OF COST DIFFERENCES FOR LOCATION

By recapitulating at this stage some of the major economic factors and policy factors bearing on location we may gain insight into these recent developments and into future prospects.

The presumed real advantage of producing for export in less-developed countries hinges upon savings anticipated in power, transportation, and labor costs. Potential savings in power costs will be the difference between hydro costs of less-developed countries and the thermal-power costs encountered in industrialized countries. Heretofore this discrepancy has been considerable—in part because of the energy policy of the industrial countries of Europe and Japan in protecting domestic coal industries. Because of this protection, thermal-power costs have gone as high as 8 mills in some countries. In the absence of protection it is reasonable to expect that imported coal or oil would succeed in holding costs down to 5–6 mills in any area accessible by sea, provided large stations operating at base-load plant factors are used in the computation. Over the longer run some easing of energy policy in the energy-deficit industrial areas can be anticipated. In the United States thermal-power costs, of course, already are in the 4-mill range.

It seems likely that within a very few years the ceiling on power cost to the aluminum industry in industrial countries outside the United States will be set by nuclear power. Already nuclear power provides a ceiling in the under-5-mill range for any industrial country prepared to import the existing equipment (free of duties). Over time the prospect is for a ceiling more nearly in the 4-mill range and possibly lower. Thus, even for those energy-deficit industrial countries which are most protectionist on conventional fuels, the advent of nuclear power should level out power costs at a lower ceiling.

Hydropower costs in less-developed areas are still largely speculative. However, in the most optimistic cases it is rare to find a figure of under 2 mills per kwh expected, and never less than 1 mill. More commonly costs are expected to be in the 2–3-mill range or higher. Therefore, instead of figures ranging between 2 mills for hydro in a less-developed country and 8 mills for thermal in some industrial countries as has appeared to be the prospect heretofore, the industry now faces a prospective gap of only 2 to 3 mills difference in power cost between industrial countries and best sites in nonindustrial countries, or, on our assumptions, $30–$45 per ton of metal. While these remain significant differences which can have an important effect on profit margins, they are within a range that can be counterbalanced by other influences.

Transportation costs with existing relationships tend to favor the refinement of bauxite into alumina near the ore sources but, for subsequent stages of the industry, transport costs are minimized by locating nearer to markets. Thus, export smelters in less-developed countries face a small transport disadvantage over the alternative of direct shipment of alumina to market. Particular conditions of the trade run make it difficult to go much beyond this statement. In some cases, advantage may be taken of favorable opportunities for return cargo or of favorable handling costs so as to upset the relationship, but these seem likely to be the exception. If regular point-to-point runs of ingot in shipload lots could be established, it should be possible to reduce the ton-mile costs of aluminum shipments to the advantage of the less-developed countries. This possibility should not be excluded for large projects affiliated with major international firms, but it remains a hypothetical one. Under the best of circumstances, carriers hauling metal would be smaller and handling charges higher than for alumina. While the less-developed countries might diminish or eliminate the present disadvantage they would incur in shipping ingot to industrial countries, they are unlikely to gain any significant transport advantage by shipping the more refined product.

It might be supposed that labor costs would be significantly lower in the less-developed countries because of markedly lower hourly wage rates. Of course there is very little experience on this matter in the aluminum industry so far, but manning scales can be expected to be significantly higher in less-developed countries—perhaps as high as 30–50 man-hours per ton of metal compared with the existing 25–30 man-hours in Western Europe and Japan and 11–14 man-hours in the United States and French plants.[3] Moreover, while wage rates may be very low in nonindustrial countries, the total remuneration of labor often includes housing and welfare measures paid by the firm, making wage rates deceptive. Finally, the more modern plants in industrial countries have reduced potroom labor almost to the vanishing point—as low as 4–5 hours per ton of metal in some cases.[4] Such manning scales are only feasible in the case of a highly skilled and disciplined work force which could not be anticipated presently in less-developed countries. They permit modern plants in advanced countries to support hourly wage rates of 3–5 times those of less-developed countries without suffering any disadvantage in unit labor costs. Therefore, both because the labor cost in advanced plants is small and the evidence suggests that other factors erode the capacity of nonindustrial countries to take advantage of their lower hourly wage rates, it seems unlikely that unit labor cost differences will significantly favor less-developed countries; indeed labor costs are not likely to affect location very greatly.

Thus, less-developed countries fail to gain a decisive advantage

---

[3] United Nations Secretariat, *Pre-investment Data on the Aluminum Industry*, ST/ECLA/Conf.11/L.24, January 28, 1963, by Jan H. Reimers, p. 32.

[4] The new Intalco plant in Bellingham, Washington, is expected to have total labor requirements of under 10 man-hours per ton, implying potroom labor of no more than 4–5 hours per ton (*Light Metal Age*, February 1965, p. 10).

from those cost elements which are favorable to them. On the other hand, they suffer a significant disadvantage from other locational factors.

In principle there is little reason for capital costs per ton of capacity to vary among industrialized countries for plants of similar design. The major industrial centers produce the necessary equipment, have the skilled personnel required to install it, and can trade with each other readily if one should have an advantage in production of some capital items. To be sure there are economies to scale, and increased capital costs can be incurred to obtain operating economies either through savings on electricity or labor. It is interesting to note that investment needed to reduce manpower requirements has occurred not only in the United States, where labor is expensive but also in France, where it is less so. Except for differences in scale related to market conditions, there is no great technical difference among the plants built in the various industrial centers. As markets grow or trading blocs consolidate, even this discrepancy may disappear.

At the same time it is conceded that real capital costs of a plant in less-developed areas may run 10–25 per cent higher than in industrial countries. This is because costs of transporting the imported equipment for the plant raises the cost, as does the necessity to bring in construction equipment and skilled construction personnel. In addition, there may be some differences in plant design. They could be dictated by the problems of heat dissipation in a tropical climate requiring more floor space per unit of capacity, while in the case of isolated plants greater self-sufficiency is required in the machine shop. Finally, start-up expenses, while not strictly a capital cost, inevitably will include extensive training of the indigenous work force in basic industrial skills. All of this is apart from the possibility that certain infrastructure items also may need to be provided, perhaps by the firm.

A sizable smelter in the 100–200-thousand metric ton range laid down in an industrial country can be figured to cost about $700 per ton.[5] The equivalent plant in a less-developed country will bear

[5] *Ibid.*, p. 26, shows a range of $500–$850 for prebaked plants of 100,000–200,000 tons. The U.S. Senate Rampart study offers a figure of $600 (U.S. Senate Subcommittee on Public Works, *The Market for Rampart Power, Yukon River, Alaska*, 87th Cong., 2nd sess. (Washington: U.S. Government Printing Office, 1962, p. 155). Most current projects are smaller and approach closer to $1,000 per ton.

a higher initial cost. In addition, firms will require that yields be higher to compensate for the additional risk. The significance of alternative assumptions concerning these two influences can be indicated, although we do not know the internal magnitudes employed by the companies in making this reckoning. Purely as an illustration, if we assume a 20-year life of plant, 10 per cent return on investment, and full utilization of plant, then an increase of capital cost-per-ton of capacity of 10 per cent will raise the level-flow expected return on capital by $7.84 per ton of metal, while, if cost is 25 per cent higher, the resulting increase is $19.59 per ton.[6] Thus, increased costs per ton of capacity at the same expected rate of return in this case would add to costs the equivalent of from 0.52 to 1.31 mills per kwh in power costs. Obviously the consequences in added cost-per-ton of metal are still greater if a shorter depreciation period is allowed for the less-developed country.

Investments in less-developed countries will be expected to yield a substantially greater rate of return to compensate for the added risk. Again as an illustration, we might choose an expected gross rate of return 5 points higher for the less-developed country (15 per cent, for example) as a basis for calculation. This acts in two ways. The first is simply to magnify the effect of a higher cost-per-ton of capacity. If we reckon with a 15 per cent gross rate of return,

[6] While only illustrative, these figures have some plausibility. Peck suggests a 20–30 year life for a smelter and figures the average return on investment in the U.S. aluminum industry at 14.8 per cent for the years 1950-55 (Merton J. Peck, *Competition in the Aluminum Industry* [Cambridge: Harvard University Press, 1961], pp. 144, 153). The return in this case is net income plus interest on long-term debt as a per cent of net worth plus long-term debt. Elsewhere (p. 60) Peck quotes an Alcoa president as seeking prices that will return a far more modest after-tax profit of 10 per cent on equity. Both of these statements pertain to a period of over 10 years ago. Today the figure might be lower. If we assume that half of the capital is borrowed at 5 per cent and pretax earnings amount to 25 per cent on the equity half (leaving about 12–13 per cent as the after-tax return on equity), then we arrive at a figure of about 15 per cent gross return on the investment. A 10 per cent gross return on investment with the same kind of leverage used in the example above allows for approximately 7.5 per cent after-tax return on equity. This comports well with the 7.2 per cent average return on assets in manufacturing reported by George Stigler over the period 1938-56 (*Capital and Rates of Return in Manufacturing Industries* [New York: National Bureau of Economic Research, 1963], p. 34).

In practice, full utilization of plant cannot be assumed. Adjustment for a utilization rate of 85 per cent would raise capital cost-per-ton of metal produced by 17.7 per cent, with attendant proportional effects throughout the computation.

then a 10 per cent increase in cost-per-ton of capacity raises the required level flow of return on capital by $10.42 per ton of metal while 25 per cent higher costs raises this item by $26.05 of the equivalent of 0.69 to 1.76 mills per kwh in power costs—a significant amount. The second and more important effect is to raise the expected rate of return directly, whatever may be the cost-per-ton of capacity. A 5 per cent increase in expected rate of return at a cost of $700 per ton of capacity means a difference of $25.83 per ton of metal.

Since the industrial country in fact is likely to benefit from both the lower cost per unit of capacity and the lower expected return on capital, the comparison is still more unfavorable for the less-developed country. In our illustration, the combined higher cost of plant and greater return on capital may require a level flow of $36.25–$51.88 per ton of metal higher in the less-developed country, which is the equivalent of about 2.4–3.5 mills per kwh in power cost (Table 23).

Table 23. Illustrative Table Showing Effect on Cost per Annual Ton of Capacity of Variations in Rate of Return and Cost of Plant[a]

Cost in Dollars of Plant per Annual Ton of Capacity	Cost in Dollars per Annual Ton of Capacity at Expected Rate of Return				
	7%	10%	12%	15%	25%
700	63.86	78.37	88.51	104.20	158.01
770 (= 10% higher)	70.25	86.21	97.36	114.62	173.81
805 (= 15% higher)	73.44	90.13	101.78	119.83	181.72
840 (= 20% higher)	76.64	94.04	106.21	125.04	189.62
875 (= 25% higher)	79.83	97.96	110.63	130.25	197.52

[a] The figures are the required annual sums flowing uniformly for 20-year periods with continuous compounding required to yield the rate of return shown. See Eugene L. Grant and W. Grant Ireson, *Principles of Engineering Economy* (4th ed.; New York: The Ronald Press Co., 1960), pp. 565–66.

To the extent that advances in technology permit savings in capital costs, the disadvantage of the higher costs in less-developed countries is diminished. Such savings may amount to 15–25 per cent over the period of concern with existing technology—possibly more if direct reduction processes are used. However, direct reduction will remain a highly technical operation, carrying costs of training and supervision in less-developed countries above those of present technology. Thus, while it might ease the disadvantage of less-developed countries, the improvement will not be entirely proportional to the lowered capital costs.

Any developments favoring lower capital costs-per-ton of capacity will be to the advantage of the less-developed countries because this will diminish the importance of the higher cost of plant in such countries. On the other hand, to the extent that developments reduce power requirements per ton of metal or raise the level of technical competence required, the nonindustrial countries become less attractive sites. Direct reduction, while a highly technical process, also offers transport advantages to those smelters located near the ore source. By preventing the industrial countries from taking advantage of their present opportunity to transport alumina instead of bauxite, it shifts the transport factor against them. However, the instances of coincidence of power and ore at the same site are few. The costs of handling are such a sufficiently large share of transport costs that, if ore must be shipped even a moderate distance for refinement, much of the advantage on transport costs which less-developed countries can aspire to with the direct-reduction process would be lost. On balance, transport is likely to be a fairly neutral element in smelter-location decisions even with the adoption of direct reduction. Over all, while technical developments seem apt to improve the position of less-developed countries vis-à-vis developed countries in several ways, the total effect is not decisive.

In addition to the foregoing elements of cost differences, most of them grounded in some sort of real economic factors, other considerations of policy and prejudice affect location. The most obvious of these is trade barriers. At present few industrial countries impose restrictions other than tariffs on trade in ingot. But if tariffs are maintained at rates of 5–9 per cent in major industrial countries (13 per cent in Japan) this is the equivalent of 2–3 mills in power cost (nearer to 5 mills for Japan)—by itself sufficient to compensate for most or all of the power cost advantage which might be anticipated in less-developed countries.

As we have suggested, the growing internationalization of the companies favors the reduction of tariffs, but there will be strong resistance as well. The resistance will come from those who have a vested interest in their home markets and who do not feel well positioned to operate on a world scale. Appeal will be made to the traditional importance of aluminum as a defense material. Other appeals will be made to the presumed advantages in service from dealing with a home-based supplier.[7] The pressure of nonindustrial countries for freer access for the products of their infant manu-facturing industries in the markets of developed countries may coincide in this instance with the desires of major international firms that favor greater flexibility. Herein lies the greatest hope for export industry in less-developed countries.

On the basis of present evidence, many European producers will seek to maintain a tariff in excess of the current American level. Until the tradition of national industries is eroded there and European-based international firms become much stronger, this degree of protection is likely to prevail and carry with it the maintenance of American tariffs near their present figure. While this will tend to inhibit the flow of outside metal to the major industrial markets, associated overseas territories of major trading blocs will enjoy favored status. It also is possible that controlled amounts of metal may be admitted under tariff quotas in response to pressure from international companies or less-developed countries.

By way of summary, two major disadvantages faced by less-developed countries that aspire to sell in the markets of industrial countries are: (1) the higher capital item resulting from greater real costs-per-unit of capacity and higher evaluation of risk and (2) tariffs encountered in industrial countries. Together these factors are sufficient to swamp prospective advantages in power cost. Moreover, the comparison is slightly worsened if transporta-tion cost is included. The effect of the two principal factors is illustrated in Table 24, allowing for three different tariff rates, two assumptions about differences in cost of plant, and a 5 per cent difference in anticipated rate of return. (See Table 23 for the basis of these figures.)

The table is hypothetical except for the tariff item. Under some conditions, assumptions B or C would not apply with full force—especially if the comparison is made with a country like Japan where domestic interest rates are high.

---

[7] It is difficult to see the merit of this argument at the ingot stage, yet it seems to be an article of faith with many in the industry. Adequate invento-ries and sufficient flexibility in international operations should provide neces-sary assurance to customers. In part this view may be a carry-over from the days of essentially national monopolies in the industry.

Table 24. Example of Differential Effect of Various Tariff Rates and Capital Charges on Cost-per-Ton of Metal

	5+% (U.S.)		Tariff Rate 9%		15%	
	Dollars	Equivalent in mills per kwh[a]	Dollars	Equivalent in mills per kwh	Dollars	Equivalent in mills per kwh
A Tariff (price $540 per metric ton)	27.55	(1.8)	48.60	(3.2)	81.00	(5.4)
B Assumption: 10% higher plant cost, 5% higher rate of return	36.25	(2.4)	36.25	(2.4)	36.25	(2.4)
C Assumption: 25% higher plant cost, 5% higher rate of return	51.88	(3.5)	51.88	(3.5)	51.88	(3.5)
A + B	63.80	(4.3)	84.85	(5.6)	117.25	(7.8)
A + C	79.43	(5.3)	100.48	(6.7)	132.88	(8.9)

[a] Based on 15,000 kwh per metric ton.

It has been speculated that over time industrial countries may reduce their tariffs, while some nonindustrial countries will enjoy favored status with major trading blocs. Also, as less-developed countries gain facilities and experience with industry, it should be possible to moderate their higher costs of laying down a plant. To the extent that such countries can offer a more secure investment climate, they will be able to reduce the margin in expected rate of return from which they presently suffer. Moreover, by combination of international financing and investment guarantees available from some industrial countries, the risk of such investment to the firm can be sharply reduced, as was the case in Ghana. A combination of these developments could make feasible some projects in less-developed countries which under present circumstances are not.

It is conceded that any country able to provide a market sufficient for a minimum-sized smelter and willing to protect it from outside competition can have a national smelting industry serving its domestic needs. Commonly such an industry can be built in collaboration with major international firms that may provide financial assistance. It will permit exchange savings, though it may involve higher-priced metal. This pattern seems likely to prevail, whether economically rational or not.

Finally, there remains the possibility that less-developed countries with good power sites may attempt to develop an export industry without affiliation with major firms. It would be possible in such cases to acquire necessary technical knowledge, and they might aspire to international financing on favorable terms so as to overcome a major disadvantage from that source. There would remain the tariff barrier and a slight transport disadvantage, which in combination would be likely to swamp the advantage of the less-developed country. Beyond this there is the serious problem of disposing of the metal in markets where most fabricating capacity is the captive of other firms. The less-developed country would be at a severe disadvantage in securing outlets, offering technical services, and providing the assurance that they were a secure source of supply. Acquisition of this marketing apparatus and sophistication would come slowly at best and is a major bar to such an independent venture.

Thus, the picture that emerges is one in which major consumption and production remain concentrated in industrial countries, contingent upon the timely provision of nuclear energy in the 5-mill range or the softening of energy-import policies to permit conventional power production at little more cost. In developed countries costs will tend to equalize based on expansion with thermal power. Less-developed countries will produce for national markets if they are large enough and may gain some entree into certain export markets if they can secure tariff relief, favorable financing, and the technical and marketing cooperation of a major international firm.

NATIONAL INTEREST AND THE ALUMINUM INDUSTRY

As we have seen, the location of the aluminum industry in the past has been influenced by the strategic considerations of governments as well as by the business calculations of private firms. Governments have been in a position to provide incentives or impose conditions which, if desired, could override private calculations. In developed countries this governmental power is less apt to be exercised in the future as the strategic importance of aluminum diminishes and habits of freer trade become established. Private firms still may seek to enlist government aid to preserve or extend protected positions, but the earlier public rationale for this has weakened and in many cases the interest of the firms also favors freer trade, with its corollary of greater locational flexibility.

Diminished governmental support in developed countries does not necessarily imply a wholesale migration of the industry. Private firms will remain alert to long-run considerations of security of investment in unstable areas and will put some price on this factor as well as on the presumed advantage to be had in producing nearer to the market. Nonetheless, under certain circumstances of appropriate investment guarantees, declining tariffs, and attractive cost advantages they will be interested in production abroad once they are less likely to encounter government-sponsored inducements to expand production at home.

In less-developed countries the attitudes of governments also are important for, while they do not control major markets, they often have control over resources needed by the industry and they may have aspirations to produce metal locally for a variety of reasons. It is worthwhile to look briefly at some of the considerations which appear to affect the attitude of governments toward the aluminum industry, both in developed and underdeveloped countries.

In developed countries governments are subject to pressures from domestic interest groups and hence are sensitive to the expressed views of the industry. National firms will continue to seek tariff protection. However, growing internationalization of the industry can be expected to make these pressures less acute insofar as tariffs and trade restrictions are concerned. Domestic producers will continue to seek favorable power cost by way of low rates from state firms, subsidized power, or concession rights to develop power at desirable sites; the outcome of these efforts will depend on the political process.

The rationale from the standpoint of the public interest for favoring domestic production would seem to be chiefly the usual autarchical arguments of avoiding dependence on foreign sources in case of emergency. This argument has less merit in a world where long wars of attrition are improbable, where free-world countries are closely linked, and in an industry where in any case there remains dependence on foreign sources of raw materials.

Great Britain is the outstanding example of a country that departs from a self-sufficiency position. The British impose no tariff on ingot yet maintain a substantial fabricating industry for which they are able to obtain imported metal at competitive prices. Japan is an example of a country where opposite views appear to retain much force. The Japanese continue to expand their domestic smelting industry even though they must import the ore, and in the future must turn to imported fuels for energy. Other countries fall in between. However justified, governments of many developed countries are likely to protect or encourage domestic industries to a moderate but diminishing extent during the period of concern.

The more interesting range of questions concerns the objectives which governments of less-developed countries seek in fostering a domestic industry. These objectives often are not made explicit. There is a general commitment to the proposition that industrialization is to be sought, and aluminum smelters are viewed as one facet of this. In other cases there is a belief that a nation is "exploited" by the sale of raw products, or that domestic resources should be further processed locally so as to add to the export value of the country's products. The country may be under foreign-exchange pressure and see a smelter as a means of easing the pressure either

240

through import substitution or via positive export earnings. In some instances the aluminum smelter is very nearly a by-product of river development plans being pursued with broader aims. In other cases a smelter may be seen as the logical means of realizing a return on hydropower resources which otherwise must remain unutilized.

Where the objective is to conserve foreign exchange currently being spent on aluminum imports, it is clear enough that a domestic smelter will serve that purpose in many cases. Since some local labor and materials also will be employed in production, imported materials and service on capital imports occasioned by the smelter ordinarily will require less exchange than importing an equivalent amount of metal. The result may be more expensive metal with consequent damage both to consumption and metal-using exports, and it may cause the diversion of other domestic resources which could have been used to earn exchange. These possible offsets must be borne in mind; but a moderately well-endowed country with a market large enough to sustain a minimal-sized plant could expect to achieve exchange savings.

Whether a less-developed country should view aluminum as a positive export-earner depends upon a number of factors. Assuming that foreign marketing problems can be solved (probably through association with a major firm) and that metal can be produced and delivered at a competitive price, it then becomes a matter of deciding whether this is the most efficient way of earning exchange or whether it entails other advantages that make it attractive.

In part this depends on the source of the funds. If a country considers investing its own resources in a smelter, it needs to consider whether this presents the best opportunity open to it to earn exchange (or to pursue other goals established). By contrast, an outside firm might only look at the problem from the standpoint of where it could most profitably locate a plant; therefore, it might be willing to make this specific investment for private reasons even though from the standpoint of the host country's national interest in earning exchange, other investments would be more advantageous. Assuming that capital is the major limiting factor on growth and that this investment would not be available for the preferred alternative, the country is well advised to allow the smelter to go forward.

Frequently, however, the decision will not be so clear-cut. While

241

an outside firm may be willing to invest in the smelter, the host country might have to make needed investment in power and infrastructure. These funds, too, may be available through international channels, but lenders will be conscious of the total foreign debt incurred and may be influenced to ration other investments. Even the direct investment in the smelter by an outside firm may be contingent on some form of investment guarantee which is subject to a form of rationing and likely to affect other projects.

For these reasons a projected smelter aimed at earning or conserving exchange should be considered in the light of some of these less obvious effects which the investment will have on exchange earnings and on other projects. Willingness of an outside firm to build an export smelter without requiring any subsidy on power price suggests that it probably is a good export earner, but one oriented to the domestic market and defended by trade barriers requires far more careful consideration.

In this connection it is useful to note the arrangements made for export smelter projects in less-developed countries, now either in operation or well advanced. Of these the French plant in Cameroon antedates the independence of the country and is therefore of less interest. The power and infrastructure investment was undertaken by the colonial government and power was made available to the company at a very favorable rate said to be around 1.5 mills per kwh.[8]

For Kaiser's Ghana project (Reynolds has a 10 per cent participation), the investment in power and infrastructure is financed by the Ghanaian government with the help of international loans to cover foreign exchange costs. The company will enjoy a 10-year tax-free operation and full control, except that it must buy an agreed amount of power at a fixed price. The smelter will be financed without any investment exposure on the part of the company; Export–Import Bank loans provide $96 million, and the $32 million equity funds are fully guaranteed by the U.S. Development Loan Fund.[9]

Alcoa's project in Surinam presents a very different picture, for in this case the company is subject to some leverage from one of its major bauxite suppliers. However, the area has provided a stable

[8] *Revue de l'aluminium*, "L'énergie, facteur essentiel pour l'industrie de l'aluminium," December 1963 (Paris), pp. 1235-37.

[9] *Fortune*, "Edgar Kaiser's Gamble in Africa," November 1961.

investment climate and, in agreeing to build alumina and smelting facilities, Alcoa was able, as part of the same package, to secure certain tax privileges, detailed guarantees for their operation, and at the same time has extended its bauxite concessions.[10]

As of this writing no final arrangements have been announced governing the proposed Guayana smelter in Venezuela. However, a regional development authority in this case will provide power and infrastructure investment while the smelter apparently will be a joint venture between the Venezuelan development bank and Reynolds, each of them putting up 20 per cent of the cost, with the remainder covered by Export–Import Bank and AID loans. The Reynolds equity share in turn will enjoy a U.S. government guarantee and the company will have tax and tariff concessions as well.[11]

In sum, these projects frequently require the less-developed country to arrange at least part of the financing (probably for power), and incentives beyond those of straight commercial venture typically have been involved.

A widespread desire to engage in further processing of domestic raw materials or a more general wish to develop idle natural resources is found in less-developed countries—indeed, this aim is not restricted to them. In part this reflects a belief that terms of trade are adverse to raw-materials suppliers or that foreign buyers are appropriating an economic rent which the country might claim. It also may be related to a planner's belief in industrialization as a long-range objective to be sought even where short-run advantages are not evident. Finally, it may be simply part of a program to industrialize so as to provide additional employment opportunities. An aluminum industry can respond to some of these aims, although it may not be the best response.

Countries that feel squeezed by worsening terms of trade for raw materials should consider the full range of possibilities open to them and not focus exclusively on the further processing of resource-based exports. Their comparative advantage well may lie in other fields rather than in such capital-intensive industries as aluminum smelting.

[10] For details of this arrangement, see Suriname-Suralco, *Brokopondo, Joint Venture.*

[11] Corporación Venezalano de Guayana, *The Guayana Region: A Portfolio of Investment Opportunities* (1963), pp. 35-36.

The desire on the part of developing countries to claim any economic rent arising through the exploitation of their resources is entirely understandable. However, it is doubtful that there is much rent to be claimed in the aluminum industry. Bauxite supplies are so large and widespread that they have limited scarcity value. In some instances there may be advantages associated with location, but presumably a country could realize them through sale of bauxite without constructing a smelter. A hydropower resource differs in that ordinarily it must be used near the site. Thus, if it is an economical site, an aluminum smelter may prove to be the only way of realizing any return on this resource.

Alumina production is a straight industrial process which provides little opportunity for economic rent. There is suspicion that alumina sold on the market is well above cost, but the major smelters produce their own or can negotiate favorable terms. It is unlikely that a developing country would find the independent production of alumina highly profitable. Of course, bauxite-producing countries welcome any additional processing they can obtain and they continue to exert pressure on the major firms to do this processing locally wherever possible.

Insofar as a country's aim in industrializing is to raise the productivity of the economy, aluminum smelting may serve, but the same investment could prove more effective in other industries. If the focus is on provision of employment opportunities, it is quickly apparent that an aluminum smelter employs relatively little labor. Stages prior to the ingot stage likewise mean little employment. Therefore a smelter or an integrated industry is neither very useful in absorbing excess labor force nor is it very likely to draw much labor away from other activities. Where the aim is to gain industrial experience or training it again is apparent that the technology employed is not broadly applicable elsewhere in the economy, so its value as a training ground is limited. Perhaps the machine shop necessary for maintenance is the most promising aspect so far as a general industrial training is concerned. In truth, the smelter is apt to remain something of an enclave rather than serve as the base for a broad range of other activities.

Yet another rationale may occur where an aluminum smelter is an integral part of a multipurpose river-development project providing the necessary market for power which permits the project to be developed on such scale that other benefits in the form of navigation, irrigation, or electricity can be made available to the economy at low cost.

In some instances, aluminum smelting clearly represents the best use that can be made of a country's resources, especially if it does not divert investment from other uses. Iceland and Norway, although advanced countries, are interesting examples of countries short of other resources but rich in power potential and able to attract foreign investment. Under these circumstances the location of smelters is a rational response to the possibilities open to them, but it is a fairly straightforward commercial proposition and not related to broader and perhaps vaguer objectives.

Apart from questions of prestige then, an aluminum smelting industry is not essential to a country for reasons of defense, industrial progress, or employment. It may help conserve foreign exchange through import substitution, but usually at the cost of more expensive metal; it may earn exchange with only a minimal diversion of domestic resources if it is commercially viable and able to attract foreign capital; or it may permit collateral benefits from river development projects.

For developed countries this discussion seems to imply that the national interest will be in ready access to the cheapest sources of supply, whether domestic or foreign. However, the private firms which supply the markets of developed countries will seek their supplies outside the developed countries only on the condition that the cost remains low after a substantial allowance has been made for security of investments made outside the developed countries. For this reason the prospects for the industry in developed countries, including some of the less densely-settled ones, remain strong. Investments can be made with security in Norway, Iceland, Canada, New Zealand, and Australia, as well as in Alaska, and metal from these sources can be assumed to be available to all non-Communist countries. However, the pull of markets and the advent of low-cost nuclear power may in fact result in a tendency for metal to be produced in the major consuming countries or trading blocs. So far as strategic considerations have any bearing, the non-Communist countries as a trading system may wish to ensure the adequacy of their supplies for strategic reasons, but this need not be done on a country-by-country basis.

For a developing country that has a market, whether domestic or foreign, the question of national interest is more complex. If it is proposed to build a smelter serving the domestic market as an aid to industrialization or to conserve exchange, then the result is apt to be more expensive metal because commonly the size of plant is small and the market will be protected by a tariff. It already has been argued that the broader effects on industrialization of a domestic smelter will be few since employment is small and the technology narrow. Very possibly far more employment and industrial skill could be developed by building other industries (fabricating, for example) with the same capital and using imported metal for domestic needs. Often, however, both in the case of a smelter and a fabricating plant, foreign capital is available which could not be had for other purposes. Moreover, the promise of a protected market (and the corollary threat of exclusion of imports) may induce this foreign capital to enter. If the effect on metal price is nil or small then it should be of benefit to the host country. If the host country must use its own general borrowing power to build a high-cost smelter, the venture then becomes more dubious.

An export smelter is presumed to be competitive in cost. Conceivably a government might subsidize such a project but, if consideration is given to the effect on domestic employment and technological throw-off, there are likely to be better alternative means of earning exchange. If domestic resources are used, the alternatives to this investment must be considered but, if foreign specialized funds are available, construction of the project may be then advisable.

CONCLUSIONS ON TRADE AND INVESTMENT

Beyond considerations of advisability, what are the real prospects for trade and investment in aluminum? It is best to focus on primary ingot because the largest trade volume will continue to be done at this stage and generalizations are more easily formulated here.

In general it appears that the nonindustrial countries of Asia, Africa, and Latin America, acting under the impetus of economic nationalism and with a bias in favor of industrialization, are seeking to supply their own needs as soon as the market is sufficient to justify even a minimum-sized plant. They may continue to do so in many cases even though the metal may be high cost and the project illogical on strictly economic grounds. The formation of regional trade groupings such as LAFTA or a potential similar organization in Africa may be expected to accelerate the attempt to supply local needs on a regional basis. To the degree that such groupings are formed and the market broadened, the likelihood of finding suitable sites and of building to economic scale is enhanced and the economic illogic diminished. Major international firms will feel compelled to participate in these endeavors so as not to be foreclosed from the markets, though they often would prefer to serve them from outside. They will form consortia and seek local participation and guarantees and reluctantly will build local plants at the expense of trade and often in defiance of the logic of locational economics.

Less-developed countries also may aspire to export markets in aluminum as a means of utilizing local power resources and in order to earn foreign exchange. In this case the projects ordinarily must meet more stringent economic tests and will require the participation of a major international firm. A major obstacle to realization of economically sound projects of this sort is the political situation in the less-developed countries which may be considered inhospitable to investment by the major firms. Given a good investment climate and very favorable cost conditions, some projects of this sort will be built. Their chances are best if they enjoy associate status with EEC or like groupings or if they can aim at such tariff-free markets as the United Kingdom. Any general reduction in tariffs or international agreements to favor the industrial exports of less-developed countries likewise would enhance their prospects.

Europe is presently a deficit area, particularly in the United Kingdom and most EEC countries. There are prospects for a substantial increase in production in Norway. Elsewhere much depends on developments in nuclear power. If we exclude nuclear power, limitations on power availability suggest that any attempt to achieve self-sufficiency would involve relatively high costs. Certainly no such effort would be in prospect for the United Kingdom. It is less certain whether the effort would be made for the EEC countries. Economical expansion is difficult for all of them except for a proposed Dutch plant built on the Groningen gas field. France might pursue a policy of self-sufficiency for national-

istic reasons but is unlikely to be able to supply the needs of other
EEC countries. Potential deficits hence would be made up in the
first instance by Norwegian and Canadian production. Thereafter
the potential of associated countries where power and bauxite are
available could be developed as political conditions permit. Obviously, expansion of the EEC to include EFTA countries would
give Norway a stronger position at the expense of Canada. European consortia could be expected to take the initiative in building
facilities in associated countries and in assuming responsibility for
marketing the product, although development in such countries
would provide better opportunity for North American firms to
penetrate the EEC market also. Under these circumstances European dependence on physically distant sources of supply could be
expected to increase, but control of the supply might remain
largely with European interests. There would seem only a very remote possibility of new private or state firms springing up in nonindustrial countries to supply the European deficit.

The big question mark concerns the possibilities for atomic
energy. It is not unreasonable to expect abundant atomic power
in the 4–5-mill-per-kwh range during the period of concern. If
this occurs, and barring any dramatic improvement in the investment climate in less-developed countries, the EEC countries are
likely to expand their domestic production to meet most local
needs. The advantages of home control over investment and better
service to customers resulting from proximity to market would be
likely to outweigh possibly cheaper power elsewhere. Economical
nuclear energy would offer a new opportunity for the United Kingdom and possibly Belgium to build smelting capacity.

The growing American market can be satisfied through increased
domestic production without great cost penalty, in contrast with
the present situation in Europe. However, given a favorable investment climate, it is possible that production abroad would prove
still more economical, and the relatively low U.S. tariff allows
foreign-production access to the United States. By building plants
abroad to serve the United States, American firms would have to
depart from the actual or potential protection of the American
tariff barrier and expose themselves to greater competition from
Canadian metal or from the potential output of international consortia in underdeveloped countries. The American industry in fact
seems headed in the direction of the widest internationalization,

but so far American investment abroad seems aimed mostly at supplying foreign markets. Some of the metal produced by American firms abroad will find its way into the United States, but the reasonable competitiveness of U.S. plants will minimize this. The low U.S. tariff allows for a mixed lot of domestic expansion and foreign supply, with most U.S. consumption continuing to be supplied by domestic plants.

Japan presents a special case. Heretofore the Japanese have maintained a policy of self-sufficiency in ingot well protected by trade barriers. The possibilities of economical expansion for a booming Japanese market are not bright with conventional energy sources. On economic grounds it would seem plausible to expect the Japanese to seek outside sources of supply. Conceivably they might become straight importers of Canadian or Australian metal. However, the more familiar pattern is for the Japanese to prefer control or joint ownership of their sources of supply and to prefer domestic production even if it involves some cost penalty. Recent adherence to GATT and OECD complicates the problem for Japan and may compel it to adopt a more open attitude. Unless it can reduce power costs, this development might force Japan to consider a wide variety of possibilities in the Pacific basin ranging from Australia and Southeast Asia to Alaska, and would make some affiliations abroad in the production of metal for the Japanese market seem a somewhat more likely possibility. However, in Japan, as in Europe, the development of cheap atomic power should revitalize prospects for domestic expansion, and the Japanese would seem most likely to seek to supply their own needs under such circumstances.

INDEX

Aardal og Sunndal Verk, 111, 114

Adelman, M. A., 194n

Africa: bauxite production and reserves, 41, 146, 147t; consumption, 16t, 20t, 24t, 72, 73t; EHV potential, 224; hydro potential, 203, 206t, 212-15; industry structure, 113; Kaiser activities, 106; primary metal production, 34, 35t, 36t; smelting capacity, 226t, 228t, 229

Alaska: hydropower potential, 203, 208-9; investment climate, 245

Alcan Aluminium Ltd., 44, 101-3, 112n, 138n, and trade policies, 134; investment climate, 153; market competition, 116; reduction technique, 169-70; role in world ingot price stabilization, 115; transportation problem, 158

Algeria: natural gas reserves, 190; smelting capacity, planned, 226t

Alnor project, 112

Alumina: bauxite requirement per ton of, 151; consumption per kg of metal, 92; costs, 88, 150-53, 156, 159t, 160; investment climate, effect on plant location, 153; plants, size of, 89; production, 43-49, 86-89, 151-52; source of, 83; supplies of smaller firms, 109; technological improvement prospects, 164; transportation costs, 156, 159t, 160

Aluminum Association, 31n, 61n, 96n, 135n

Aluminum Company of America (Alcoa), 103-4; 208n, 233n: investment climate and alumina

plant location, 153; Surinam project, 242-43

Aluminum industry: competition, 114-17; basic technology, 83-97; cost structure, 145-68; international, structure of, 6, 8-9, 99-117, 172; national interests and, 239-46

Aluminum-using industries: prospects, 66-70

Alusuisse. See Schweitzerisches Aluminum A. G.

Americas: consumption, 16t, 24t, 25; primary metal production, 34, 35t

Anaconda Aluminum Company: 110; use of bauxite substitute, 167, 168

Angola: hydro potential, 206t, 212, 213; smelting capacity, planned, 226t

Anodes: "anode effect", 91; cost, 154

Arbeitsgemeinschaft deutscher wirtschaftwissenschaftlicher Forschungsinstitut, 193

Argentina: Alcan operations, 102; coal reserves, 191t; hydro potential, 205t; Kaiser activities, 106; oil reserves, 190; smelting capacity, planned, 227t; tariffs, 130t

Asia: bauxite and lower quality ore reserves, 148t; consumption, 16t, 19, 24t, 25, 72, 73t; EHV potential, 224; hydro potential, 206t, 216-17; primary metal, production and demand, 34, 35t, 228t; smelting capacity, 226t, 228t, 229

Aswan Dam, 213

Australia: bauxite production and reserves, 41, 106, 146, 148t; coal reserves, 190, 191t; fuel costs, 192,

251

Metallgesellschaft Aktiengesellschaft, 22n, 41n, 43n, 44n, 49n, 75n, 77n, 120n, 121n, 141, 142n
Methodology: demand forecasting, 53-58; regional consumption projections, 78-81
Mexico: Alcoa activities in, 104; consumption, 20t, 26t; hydro potential, 205t; oil reserves, 190; production, primary metal, 38t; smelting, 111, 139, 226t; trade 124t
Middle East: oil and natural gas reserves, 190
Milling, prospects, 64
Minerals Yearbook, 151n
Mineraux et Métaux, 42n
Mining. See Bauxite, mining and processing
Monohydrate plant: investment, 151; operating factors, 88-89
Montecatini, Soc. Generale per l'Industria Mineraria e Chimica, 112
Mossé, R., 33n, 53n, 63

National Association of River and Harbor Contractors, 157n
National Coal Association, 186n
National interests, and the aluminum industry, 239-46
Natural gas reserves, 190-92
Nepheline, as substitute for bauxite, 167
Netherlands: Alcan operations, 102; Alusuisse activities, 108; coal reserves, 190, 191t; consumption, 27t, 30t; expansion prospect, 247; gas-based industry, 191-92; natural gas reserves, 190; smelting capacity, 226t; tariffs, 130t; trade, 123t, 125t; trade policy, 136
New Guinea: hydro potential, 203, 207t, 217
New York State Power Authority, 179, 208
New Zealand: consumption, 28t; hydro potential, 203, 207t, 217; investment climate, 245; low quality ore reserves, 148t; smelting capacity, planned, 227t; tariffs, 130t; trade, 127t
Nigeria: hydro potential, 206t, 212; trade, 127t
Nitride process, 171

Nonbauxite materials, technological prospects, 166-67
Norsk Hydro, 112
North America: alumina production, 89; bauxite and lower quality ores, reserves, 147t; consumption, 16t, 72; hydro potential, 204-10, 205t; labor requirement, 93; major international companies, 100-1; plant costs, 93; primary metal production, 34; smelting capacity, 226t, 227, 229
North American Coal Company, 167n
Norway: alumina production, 43; consumption, 27t, 30t; hydro, available and potential, 203, 207t, 218; international companies in, 102-10 passim; investment climate and prospect, 245, 247; position in international market, 111-12; power cost, 219; production, 38t, 40t, 43; smelting capacity, 226t, 227; tariffs, 130t; trade, 122t, 125t, 247; trade policies and company attitudes, 135-36
Nuclear energy: capital costs, 197-99; cost per kwh, 196, 199, 200; fuel costs, 196, 198; prospects for, 195-200; and trade and investment prospects, 248, 249
Nuclear Industry, 197n, 198n

Oceania: consumption, 16t, 19, 20t, 24t, 72, 73t; hydro potential, 207t, 217-18; production, primary metal, 36t; smelting capacity, 227t, 228t
Ohio Valley: coal-based plants, proposed, 104, 173
Oil, reserves and prices, 190-91, 194
Olin-Mathieson, 44, 167
Organisation for Economic Cooperation and Development (OECD), 29n, 31n, 33n, 138, 249
Oriental Economist, 155n
Ormet Corporation, 109-10
Oyster Creek nuclear station, 182

Pakistan: hydro potential, 206t, 216; natural gas reserves, 190; tariffs, 130t
Pearson, T. G., 92n, 168n
Péchiney Compagnie de Produits Chimique et Électrométallurgiques,